Observing interaction:
An introduction to
sequential analysis

Second edition

ROGER BAKEMAN
Georgia State University

JOHN M. GOTTMAN
University of Washington

CAMBRIDGE
UNIVERSITY PRESS

CAMBRIDGE UNIVERSITY PRESS
Cambridge, New York, Melbourne, Madrid, Cape Town, Singapore, São Paulo

Cambridge University Press
The Edinburgh Building, Cambridge CB2 2RU, UK

Published in the United States of America by Cambridge University Press, New York

www.cambridge.org
Information on this title: www.cambridge.org/9780521450089

© Cambridge University Press 1997

First published 1986
Second edition 1997

A catalogue record for this publication is available from the British Library

Library of Congress Cataloguing in Publication data

Bakeman, Roger.

Observing interaction: an introduction to sequential analysis /
Roger Bakeman, John M. Gottman. – 2nd ed.

p.   cm.

Includes bibliographical references and index.

ISBN 0-521-45008-X (hardcover). – ISBN 0-521-57427-7 (pbk.)

1. Social sciences – Statistical methods. 2. Sequential analysis. 3. Social
interaction – Statistical methods. I. Gottman, John Mordechai. II. Title.

H62.B284   1997   96-26084
                                                         CIP

ISBN-13 978-0-521-45008-9 hardback
ISBN-10 0-521-45008-X hardback

ISBN-13 978-0-521-57427-3 paperback
ISBN-10 0-521-57427-7 paperback

Transferred to digital printing 2006

Contents

Preface to the second edition

Since the first edition of *Observing Interaction* appeared in 1986, the technology supporting recording and systematic coding of behavior has become less expensive, more reliable, and considerably less exotic (Bakeman, in press; revised chapter 3, this volume). Cumbersome videotape and cassette recorders have given way to video camcorders. Visual time codes are routinely recorded as part of the picture, and equipment to write and read machine-readable time codes is readily available at reasonable cost. Increasingly, computers assist coding, making routine what once was labor intensive and time-consuming. Even physiological recording devices can be added to the computer's net (Gottman & Bakeman, in press). Thus an increasing circle of investigators can avail themselves of the methods detailed in this book without mortgaging their careers, their lives, or the lives of their associates.

At the same time, the way we think about sequential data has developed. This is reflected in a standard format for sequential data (Bakeman & Quera, 1992, 1995a; revised chapter 5, this volume). SDIS – the Sequential Data Interchange Standard – has greatly facilitated the analysis of sequential data. Again, an enterprise that formerly was time-consuming and cumbersome has yielded to appropriately designed computer tools, as described in my and Quera's *Analyzing Interaction* (1995), which should be regarded as a companion to this volume. This revised version of *Observing Interaction* still explains how to conceptualize, code, record, organize, and analyze sequential data, but now *Analyzing Interaction* provides the tools to do so easily.

Another area of considerable development, and one responsible for many of the differences between the first and second editions of *Observing Interaction*, concerns techniques for analyzing sequential data (chapters 7–9, this volume; these chapters are extensively modified versions of chapters 7–8 from the first edition). Formerly many of the analytic techniques proposed for sequential analysis were somewhat piecemeal and post hoc, yet, waiting in the wings, log-linear analysis promises a coherent analytic view for sequential phenomena (Bakeman & Quera, 1995b). This revised edition moves log-linear techniques (which can be thought of as a multidimensional extension of chi-square tests; see Bakeman & Robinson, 1994) center stage. This simplifies matters and, at the same time, integrates lag-sequential analysis with an established and well-supported statistical tradition.

In the preface to the first edition, I suggested that far more people deserve thanks than can be named explicitly. That is still so. Nonetheless, I would like to thank three colleagues from the University of Barcelona: Maria Teresa Anguera and Angel Blanco, who translated the first edition of *Observing Interaction* into Spanish, and Vicenç Quera, who has collaborated with me these past several years in thinking about sequential analysis and in developing the SDIS and the sequential analysis program we call the Generalized Sequential Querier (GSEQ). I would also like to thank Debora Gray, who emerged from her mountain fastness to redraw one last figure for this second edition, and all those students in my observational methods class, winter quarter 1996, who offered comments. Finally, I would like to correct an error from the first edition's preface. Mildred Parten did not disappear. She lived her life as a bureaucrat in Washington, D.C., at a time in our history when such jobs were among the few open to women who wanted both to work and to use their brains.

ROGER BAKEMAN

References

Bakeman, R. (in press). Behavioral Observations and Coding, In H. T. Reis & C. K. Judd (Eds.), *Handbook of research methods in social psychology*. New York: Cambridge University Press.

Gottman, J. M., & Bakeman, R. (in press). Systematic observational techniques or building life as an observational researcher. In D. Gilbert, S. Fiske, & G. Lindzey (Eds.), *Handbook of social psychology* (4th Ed.). New York: McGraw-Hill.

Preface to the first edition

Sometimes even a rather lengthy series of thoughts can be associated in memory with a single, dominant image. For me (RB), when I reflect on the writing of this book, the image of Mount Rainier, seemingly floating on the late afternoon haze, is never far from my mind. This is a personal image, of course, but it is understandable if I explain that John Gottman and I first met at a conference, organized by Jim Sackett, held at Lake Wilderness in Washington State. Thus some of the conversations that laid the groundwork for this book took place against the backdrop of Mount Rainier, dominating the horizon at the other end of the lake.

The conference was concerned with the Application of Observational/Ethological Methods to the study of Mental Retardation, a title that suggests both some of the research traditions that have influenced our writing and some of the kinds of readers who might find this book useful. Throughout this century, some of the most systematic and productive observers of social interaction have been ethologists, especially those concerned with primates, and developmental psychologists, especially those concerned with infants and young children.

Although students of primate and children's behavior have been largely responsible for the development of systematic observational methods, they are not the only kinds of researchers who want to study social behavior scientifically. Among others, this book should interest investigators in all branches of animal behavior, in anthropology, in education (including those concerned with classroom evaluation as well as early childhood education), in management, in nursing, and in several branches of psychology (including child, community, developmental, health, organizational, and social), as well as investigators concerned with mental retardation.

As the title implies, this book is a primer. Our intent is to provide a clear and straightforward introduction to scientific methods of observing social behavior, of interest to the general practitioner. Avoided are arcane byways of interest primarily to statistical specialists. Because the dynamics of people (and other animals) interacting unfold in time, sequential approaches to observing and understanding social behavior are emphasized.

We assume that most readers of this book will be researchers (either advanced undergraduates, graduate students, or established investigators already seasoned in other methods) who want to know more about systematic observational methods and sequential analysis. Still, we hope that those readers who are already experienced in observational methods will recognize what they do clearly described here and may even gain additional insights.

We further assume that most readers will have a rudimentary knowledge of basic statistics, but, we should emphasize, an advanced knowledge of statistical analysis is not required. Instead, throughout the book, conceptual fundamentals are emphasized and are presented along with considerable practical advice. Our hope is that readers of this book will be equipped to carry out studies that make use of systematic observational methods and sequential analysis.

Writers of acknowledgments frequently admit that far more people deserve thanks than can be explicitly named. In this, I am no exception. I am very much in debt, for example, to students and other researchers who have consulted with me regarding methodological issues, because their questions often forced me to consider problems I might not otherwise have considered. I am also in debt to those mothers, infants, and children who served as subjects in my own research, because the studies in which they participated provide many of the examples used throughout this book. Finally, I greatly appreciate those graduate students in my observational methods course who read and commented on an earlier draft of this book.

Debora Gray is responsible for many of the figures included here. I enjoyed working with her and very much appreciate her vision of what the careful students's notebook should look like. I also want to thank Melodie Burford, Connie Smith, and Anne Walters, who diligently ferreted out errors in a near-final copy of the manuscript.

Chief among the people whose encouragement and support I value is Jim Sackett. Early on, even before the Lake Wilderness conference in 1976, he urged me to pursue my interest in observational methods and sequential analysis. So did Leon Yarrow and Bob Cairns, to whom I am similarly indebted for encouraging comments at an early point in my career.

Sharon Landesman-Dwyer also helped us considerably. Along with Kathy Barnard and Jim Sackett, she arranged for us to spend a month at the University of Washington in July 1983, consulting and giving lectures. Not only did this give us a chance to discuss this book (with Mount Rainier again in the background), it gave us a chance to present some of the material incorporated here before a live, and very lively, audience.

I have been fortunate with collaborators. Most obviously, I appreciate the collaboration with John Gottman that has resulted in this book, but also I have benefited greatly from my collaboration, first with Josephine Brown, and more recently with Lauren Adamson. Their contribution to this book is, I hope, made evident by how often their names appear in the references. I would also like to thank my department and its chair, Duane M. Rumbaugh, for the support I have received over the years, as well as the NIMH, the NIH, and the NSF for grants supporting my research with J. V. Brown and with L. B. Adamson.

Finally, I would like to acknowledge the seminal work of Mildred Parten. Her research at the University of Minnesota's Institute of Child Development in the 1920s has served as a model for generations of researchers and is still a paradigmatic application of observational methods, as our first chapter indicates. How much of a model she was, she probably never knew. She did her work and then disappeared. In spite of the best efforts of Bill Hartup, who until recently was the director of the Institute, her subsequent history remains unknown.

Parten left an important legacy, however – one to which I hope readers of this book will contribute.

ROGER BAKEMAN

A small percentage of current research employs observational measurement of any sort. This is true despite the recent increased availability of new technologies such as electronic notepads and videotape recording. It may always be the case because it is more costly to observe than to use other methods such as questionnaires.

We were motivated to write this book because we believe that observational methods deserve a special role in our measurement systems. First, we think the descriptive stage of the scientific enterprise is extremely productive of research hypotheses, models, and theory. This ethological tradition is full of examples of this fact, such as Darwin's classic work on emotional expression. Second, the time is ripe for a reconsideration of observational techniques because we now know a lot more about what to observe, how to construct reliable measurement networks, and how to analyze data to detect interaction sequences. Recently, we have been making new headway on old problems with these new technologies.

We are optimistic that this book will fill a gap and stimulate new research that employs useful (and not superficial) observational systems.

I (JMG) would like to acknowledge grants MH29910, MH35997, and RSDA K200257 and sabbatical release time during 1984–1985.

JOHN M. GOTTMAN

1

Introduction

1.1 Interaction and behavior sequences

Birds courting, monkeys fighting, children playing, couples discussing, mothers and infants exchanging gleeful vocalizations all have this in common: Their interaction with others reveals itself unfolded in time. This statement should surprise no one, certainly not readers of this volume. What is surprising, however, is how often in the past few decades researchers interested in dynamic aspects of interactive behavior – in how behavior is sequenced moment to moment – have settled for static measures of interaction instead. This need not be. In fact, our aim in writing this book is to demonstrate just how simple it often is not just to record observation data in a way that preserves sequential information, but also to analyze that data in a way that makes use of – and illuminates – its sequential nature.

We assume that readers of this book may be interested in different species, observed at various ages and in diverse settings, but that most will be interested specifically in observing interactive social behavior. This is because we think sequential methods are tailor-made for the study of social interaction. As noted, a defining characteristic of interaction is that it unfolds in time. Indeed, it can hardly be thought of without reference to a time dimension. Sometimes we are concerned with actual time units – what happens in successive seconds, for example; at other times, we are just concerned with what events followed what. In either case, we think it is a sequential view that offers the best chance for illuminating dynamic processes of social interaction.

For example, we might ask a married couple to fill out a questionnaire, and from their responses we might assign them a "marital satisfaction" score. This would at least let us try to relate marital satisfaction to other aspects of the couple's life circumstances, like their own early experience or their current work commitments, but such a static measure would not tell us much about how the couple interacts with each other, or whether the way in which they interact relates in any way to how satisfied they report being with their relationship. In order to "unpack" the variable of marital

1

satisfaction, we would need to examine more closely just how the couple related to each other – and, in order to describe and ultimately attempt to understand the dynamics of how they relate to each other, a sequential view is essential. Our hope is that readers of this book not only take such a sequential view, but also will learn here how to describe effectively the sequential nature of whatever interaction they observe.

1.2 Alternatives to systematic observation

There is a second assumption we make about readers of this book. We assume that they have considered a variety of different methods of inquiry and have settled on systematic observation. For a moment, however, let us consider the alternatives. When studying humans, at least those able to read and write, researchers often use questionnaires, like the marital satisfaction inventory mentioned. These questionnaires, as well as tests of various sorts, certainly have their uses, although capturing the dynamic quality of behavior sequences is not one of their stronger points.

We do not mean to suggest, however, that investigators must choose between observational and other methods. In our own work, we usually employ a convergent measurement network that taps the constructs we are interested in studying. This network usually includes questionnaire, interview, and other measurement operations (e.g., sociometric measures). Still, we think there is something captured by observational procedures that eludes these other measurement procedures. Nonetheless, there are at least two time-honored alternatives to systematic observation for capturing something of the sequential aspect of interaction. These are (a) narrative descriptions, and (b) properly designed rating scales.

If we were forced to choose between the adjectives "humanistic" and "scientific" to describe narrative descriptions, we would have to choose humanistic. We do so, not to demean the writing of narrative reports – a process for which we have considerable respect – but simply to distinguish it from systematic observation. After all, not only have humanistic methods of inquiry been used far longer than the upstart methods we characterize as scientific, but also the preparation of narrative reports, or something akin to it, is an important part of code development, a process that must precede systematic observation.

Still, narrative reports depend mightily on the individual human doing them, and judgments about the worth of the reports are inextricably bound up with judgments about the personal qualities of their author. In fact, we would be surprised if two reports from different authors were identical. With systematic observation, on the other hand, the goal is for properly trained observers to produce identical protocols, given that they observed the same stream of behavior. The personal qualities of the observers (assuming some talent and proper training) should not matter. It is primarily this drive for replicability that we think earns systematic observation the appellation "scientific" and that distinguishes it from the writing of narratives.

Rating scales, on the other hand, allow for every bit as much replicability as systematic observation does. In fact, if we defined systematic observation more broadly than we do (see next section), the use of rating scales could easily be regarded simply as another instance of systematic observation. For present purposes, however, we prefer to regard them separately and to point out some of the differences between them. (For a discussion of some of these differences, see Cairns & Green, 1979.) Imagine that we are interested in the responsivity of two individuals to each other, for example, a husband and wife or a mother and baby. We could define behavioral codes, code the stream of behavior, and then note how and how often each was "responsive" to the other (systematic observation), or we could train observers to rate the level of responsivity that characterized the interaction observed.

For many purposes, a rating-scale approach might be preferable. For example, imagine an intervention or training study in which an investigator hopes to change maternal responsivity to infant cues. In this case, what needs to be assessed is clearly known and is a relatively coherent concept which can be clearly defined for raters. Then the far less stringent time demands of a rating-scale approach would probably make it the methodology of choice. On the other hand, if a researcher wants to describe exactly how mothers are responsive to their infants and exactly how this responsivity changes with infant development, then the more detailed methods of systematic observation are required.

1.3 Systematic observation defined

For present purposes, we define systematic observation as a particular approach to quantifying behavior. This approach typically is concerned with naturally occurring behavior observed in naturalistic contexts. The aim is to define beforehand various forms of behavior – behavioral codes – and then ask observers to record whenever behavior corresponding to the predefined codes occurs. A major concern is to train observers so that all of them will produce an essentially similar protocol, given that they have observed the same stream of behavior.

The heart and foundation of any research using systematic observation is the catalog of behavior codes developed for a particular project (see chapter 2). As inventories of questions are to personality or marital satisfaction research, as IQ tests are to cognitive development research, so are code catalogs (or coding schemes) to systematic observation. They are the measuring instruments of observational research; they specify which behavior is to be selected from the passing stream and recorded for subsequent study.

In many ways, systematic observation is not very different from other approaches to behavioral research. Here, too, investigators need to say what they hope to find out; they need to define what seems important conceptually, they need to find ways to measure those concepts, and they need to establish the reliability of their measuring instruments. However, because human observers are such an important part of the instrumentation, reliability issues loom especially large in observational research, a matter which we discuss further in chapter 4.

In sum, the twin hallmarks of systematic observation are (a) the use of predefined catalogs of behavioral codes, (b) by observers of demonstrated reliability. The entire process of defining and developing coding schemes followed by training observers to acceptable levels of agreement can be both time-consuming and demanding. But without such an effort, the investigator who goes no further than only telling others what he or she sees runs the risk of having skeptical colleagues dismiss such narrative reports as just one person's tale spinning.

1.4 A nonsequential example: Parten's study of children's play

An early and well-known example of systematic observation is Mildred Parten's (1932) study of social participation among preschool children, conducted at the University of Minnesota's Institute of Child Welfare in the late 1920s. There are in fact many excellent observational studies of children's behavior which were done in the 1920s and 1930s, and many of the basic techniques still in use were first articulated then. We discuss Parten's study here as an exemplar of that early work and as a way of defining by example what we mean by "systematic observation." At the same time, Parten's study was not sequential, as we use the term, and so describing both what she did and what she did not do should clarify what we mean by "sequential."

During the school year of 1926–1927, some 42 children whose ages ranged from not quite 2 to almost 5 years were observed during indoor free play. Parten was interested in the development of social behavior in young children, and to that end defined six levels or categories of social participation as follows:

1. *Unoccupied.* The child does not appear to be engaged with anything specific; rather, his behavior seems somewhat aimless. He

might watch something of momentary interest, play with his own body, just wander around, or perhaps stand or sit in one place.

2. *Onlooker.* The child watches other children play, but does not enter into their play. This differs from Unoccupied because the child is definitely watching particular children, not just anything that happens to be exciting.

3. *Solitary Independent Play.* The child plays alone and independently with whatever toys are of interest. The child's activity does not appear affected by what others are doing.

4. *Parallel Activity.* The child still plays independently, but his activity "naturally brings him among other children." He plays beside them, not with them, but with toys that are similar to those the children around him are using. There is no attempt to control the coming or going of children in the group.

5. *Associative Play.* The child plays with other children. There may be some sharing of play material and mild attempts to control which children are in the group. However, there is no division of labor or assigning of roles: Most children engage in essentially similar activity. Although each child acts pretty much as he or she wishes, the sense is that the child's interest lies more with the association with others than with the particular activity.

6. *Cooperative or Organized Supplementary Play.* The child plays in a group that is organized for some purpose. The purpose might be to dramatize a situation – for example, playing house – or to play a formal game, or to attain some competitive goal. There is a sense of belonging or not to the group. There is also a division of labor, a taking of roles, and an organization of activity so that the efforts of one child are supplemented by those of another. (The above definitions are paraphrased from Parten, 1932, pp. 250–251.)

Each child was observed for 1 minute each day. The order of observation was determined beforehand and was varied systematically so that the 1-minute samples for any one child would be distributed more or less evenly throughout the hour-long free-play period. On the average, children were observed about 70 different times, and each time they were observed, their degree of social participation was characterized using one of the six codes defined above.

Florence Goodenough (1928) called this the method of repeated short samples. Today it is often called "time sampling," but its purpose remains the same. A number of relatively brief, nonsuccessive time intervals are categorized, and the percentage of time intervals assigned a particular code is used to estimate the proportion of time an individual devotes to that kind of activity. For example, one 3-year-old child in Parten's study was observed 100 times. None of the 1-minute time samples was coded Unoccupied, 18 were coded Solitary, 5 Onlooking, 51 Parallel, 18 Associative, and 8 Cooperative. It seems reasonable to assume that had Parten observed this child continuously hour after hour, day after day, that about 51% of that child's time would have been spent in parallel play.

The method of repeated short samples, or time sampling, is a way of recording data, but it is only one of several different ways that could be used in an observational study. What makes Parten's study an example of systematic observation is not the recording strategy she used but the coding scheme she developed, along with her concern that observers apply that scheme reliably.

Parten was primarily concerned with describing the level of social participation among children of different ages, and with how the level of social participation was affected by children's age, IQ, and family composition. For such purposes, her coding scheme and her method of data collection were completely satisfactory. After all, for each child she could compute six percentages representing amount of time devoted to each of her six levels of social participation. Further, she could have assigned, and did assign, weights to each code (-3 to Unoccupied, -2 to Solitary, -1 to Onlooker, 1 to Parallel, 2 to Associative, and 3 to Cooperative), multiplied a child's percent scores by the corresponding weights, and summed the resulting products, which yielded a single composite social participation score for each child – scores that were then correlated with the child's age and IQ.

Knowing that older children are likely to spend a greater amount of time in associative and cooperative play than younger ones, however, does not tell us much about moment-by-moment social process or how Parten's participation codes might be sequenced in the stream of behavior. This is not because her codes are inadequate to the task, but because her way of recording data did not capture behavior sequences. There is no reason, of course, why she should have collected sequential data – her research questions did not require examining how behavior is sequenced on a moment-by-moment basis. However, there are interesting questions to ask about the sort of children's behavior Parten observed that do require a sequential view. An example of such a question is presented below.

1.5 Social process and sequential analysis

The purpose of this book is to emphasize sequential analyses of sequentially recorded data, but we should not let this emphasis obscure how useful and interesting nonsequential data (or the nonsequential analysis of sequential data) can be. At the same time, we want to argue that sequential analyses

can provide an additional level of information about whatever behavior we are observing, a level that is not accessible to nonsequential analyses.

In many ways, Parten's study typifies the sort of "time-budget" information that nonsequential analyses of observational data can provide. Indeed, it is often very useful to know how children, or mothers with infants, or animals in the wild, or office workers distribute their time among various possible activities. Nor is time the only thing that can be "distributed." We could, for example, observe married couples in conversation, code each "utterance" made, and then report percent scores for each utterance code. Computing such percentages is a nonsequential use of the data, to be sure, but it does allow us to determine, for example, whether disagreements are more common among "distressed" as opposed to "nondistressed" couples.

There are, however, additional questions that can be asked. When utterances are recorded sequentially, we can go on to ask what happens after one spouse disagrees or after one spouse complains. Are there characteristic ways the other spouse responds? Are these ways different for husbands and wives? Are they different for distressed and nondistressed couples? (For answers to these questions, see Gottman, 1979a.) At this point, we are beginning to probe social process in a way that only sequential analyses make possible.

In general, when we want to know how behavior works, or functions, within an ongoing interaction, some form of sequential analysis is probably required. For example, a nonsequential analysis could tell us that distressed husbands and wives complain more than nondistressed ones do, but only a sequential analysis could tell us that distressed couples, but not nondistressed ones, tend to react to each other's complaints with additional complaints. Similarly, a nonsequential analysis can tell us that 3-year-olds engage in less parallel play than 2-year-olds, but only a sequential analysis can tell us if, in the moment-by-moment stream of activity, young children use parallel play as a bridge into group activity. An example of such a sequential analysis will be discussed later, but first we present a second nonsequential example.

1.6 Another nonsequential example: Smith's study
of parallel play

Parten believed that her study established a relationship between children's age and their degree of participation in social groups: As children became older, they participated more. Her cross-sectional study is often interpreted as suggesting a developmental progression; thus parallel play is seen as a "stage" through which children pass as they develop from solitary to social

group players. This idea found its way into textbooks but was not tested empirically until Peter Smith did so in the late 1970s (Smith, 1978). In the present context, Smith's study is interesting for at least three reasons: for what he found out, for the way he both made use of and modified Parten's coding scheme, and for his method, which only appears sequential, as we define the term.

For simplicity, Smith reduced Parten's six categories to three:

1. *Alone,* which lumped together Parten's Unoccupied, Onlooker, and Solitary
2. *Parallel,* as defined by Parten
3. *Group,* which lumped together Parten's Associative and Cooperative

After all, because he wanted to test the notion that Parallel play characterizes an intermediate stage of social development, finer distinctions within Alone and within Group play were not necessary. Smith then used these codes and a time-sampling recording strategy to develop time-budget information for each of the 48 children in his study. However, Smith did not compute percent scores for the entire period of the study, as Parten did, but instead computed them separately for each of six successive 5-week periods (the entire study took 9 months). These percent scores were then used to code the 5-week periods: Whichever of the three participation categories occurred most frequently became the category assigned to a time period.

Smith's method is interesting, in part because it forces us to define exactly what we mean by a sequential approach. Certainly his method has in common with sequential approaches that successive "units" (in his case, 5-week periods) are categorized, that is, are matched up with one of the codes from the coding scheme. However, what Smith did does not satisfy our sense of what we usually mean by "sequential." It is only a matter of definition, of course, but for the purpose of this book we would prefer to reserve the word "sequential" for those approaches that examine the way discrete sequences of behavior occur. Normally this means that sequential approaches are concerned with the way behavior unfolds in time, as a sequence of relatively discrete events, usually on a moment-by-moment or event-by-event basis. In contrast, Smith's 5-week periods are not at all discrete, and thus his approach is not sequential – as we use the term here – but is a reasonable data reduction technique, given the question he sought to answer.

1.7 A sequential example: Bakeman and Brownlee's study of parallel play

What Smith reported is that many children moved directly from a 5-week period in which Alone play predominated, to one in which Group play

Table 1.1. *Three coding schemes for participation in social groups*

Parten (1932)	Smith (1978)	Bakeman & Brownlee (1980)
Unoccupied		Together
----------------		--------------------
		Unoccupied
Onlooker	Alone	
----------------		--------------------
Solitary		Solitary

Parallel	Parallel	Parallel

Associative		
----------------	Group	Group
Cooperative		

Note: A coding scheme is an investigator's attempt to cleave an often intractable world "at the joints." Given here are coding schemes used by the three studies discussed in this chapter. The dashed lines indicate that what Parten coded Unoccupied, Bakeman and Brownlee might have coded either Together or Unoccupied. Similarly, what Bakeman and Brownlee coded Unoccupied, Parten might have coded either Unoccupied or Onlooker. Smith would have coded all of these Alone, as well as what both Parten, and Bakeman and Brownlee, coded Solitary.

did, without an intervening period during which Parallel play was most frequent. He concluded that a period during development characterized by parallel play may be optional, a stage that children may or may not go through, instead of obligatory, as Parten seems to have suggested. Still, Smith's children engaged in parallel play about a quarter of the time, on the average, and therefore, although it was seldom the most frequent mode of play, it was nonetheless a common occurrence. This caused Bakeman and Brownlee (1980) to think that perhaps parallel play might be more fruitfully regarded, not as the hallmark of a developmental stage, but as a type of play important because of the way it is positioned in the stream of children's play behavior. Thus Bakeman and Brownlee raised a uniquely sequential question about parallel play, one quite different from the question Parten and Smith pursued.

Like Smith, Bakeman and Brownlee modified Parten's coding scheme somewhat (see Table 1.1). They defined five codes as follows:

1. *Unoccupied,* which lumped together Parten's Unoccupied and Onlooker.
2. *Solitary.* Unlike Smith, Bakeman and Brownlee chose to keep Unoccupied and Solitary separate. Because they were interested in how these "play states" are sequenced, and because both Solitary and Parallel play involve objects, whereas Unoccupied does not, they thought the distinction worth preserving.
3. *Together.* As far as we know, this code has not been used in other published studies. It appears to be a particularly social way of being unoccupied and is characterized by children clearly being with others – there seems to be an awareness of their association – but without the kind of focus on objects or activities required for Parallel or Group play.
4. *Parallel,* as defined by Parten.
5. *Group.* Like Smith, Bakeman and Brownlee lumped together Parten's Associative and Cooperative.

The source material for this study consisted of videotapes, made during indoor free play. Forty-one 3-year-olds were taped for about 100 minutes each. Observers then viewed these tapes and decided which of the five codes best characterized each successive 15-second interval. This method of recording data represents something of a compromise. It would have been more accurate if observers had simply noted when a different "play state" started. That way, not only would an accurate sequencing of states have been preserved, but accurate time-budget information (percentage of time spent in the various play states) would have been available as well.

This raises an interesting question. Is continuous recording (noting times when different codable events begin and end) better than interval recording (assigning codes to successive time intervals)? We shall have considerably more to say about this matter later. For now, let us simply say that Bakeman and Brownlee were able to extract from their data a reasonably accurate sequence of events, that is, a record of the way different play states followed each other in time.

Viewing their data as a sequence of play states, Bakeman and Brownlee first counted how often each code followed the other codes (for example, they determined how often Group followed Parallel, followed Together, etc.). Then, using methods described in chapter 7, they compared observed counts to their expected values. This was done separately for each possible transition for each child, which means, for example, that if the Parallel to Group transition occurred at greater than expected levels for a particular child, the expected levels were based on how often that child engaged in Group play.

Among other things, Bakeman and Brownlee wanted to know if certain transitions were especially characteristic of the children they observed. Of particular interest was the Parallel to Group transition, one they thought should be frequent if parallel play functions as a bridge into group play. Now just by chance alone, observed values for this transition should exceed expected values for about half of the children. In fact, observed values for the Parallel to Group transition exceeded chance for 32 of the 41 children observed, a deviation from chance that was significant at better than the .01 level (determined by a two-tailed sign test). Thus, Bakeman and Brownlee concluded, the movement from parallel to group play may be more a matter of moments than of months, and parallel play may often serve as a brief interlude during which young children have both an increased opportunity to socialize as well as a chance to "size up" those to whom they are proximal, before plunging into more involved group activity.

The point of this example is that, given a sequential view, coupled with what are really quite simple statistics, Bakeman and Brownlee were able to learn a fair amount about young children's experience in free-play groups. For one thing, it appears that children changed the focus of their activity in quite systematic ways, "one step" at a time. Some transitions were "probable," meaning that observed exceeded expected values for significantly more than half of the children. Other transitions were "improbable," meaning that observed exceeded expected values for significantly less than half of the children (for example, observed exceeded expected values for the Unoccupied to Parallel transition for only 2 of the 41 children). The remaining transitions were neither probable nor improbable; Bakeman and Brownlee called them "possible" or "chance" transitions.

The probable transitions all involved either remaining alone (moving between Unoccupied and Solitary) or else remaining with others (moving between Together, Parallel, and Group). What is interesting, however, is how any movement at all occurred between being Alone and being Together. Thus transitions from Unoccupied to Together (adding a social focus to no focus) and from Solitary to Parallel (adding a combined object and social focus to an existing object focus) were possible, whereas transitions from Unoccupied to Parallel (which would require simultaneously adding both an object and a social focus) or from Solitary to Together (which would require simultaneously dropping an object focus and adding a social one) were improbable. Theoreticians can now argue about what this "one step at a time" model means, but for present purposes, we would just like to emphasize again that without a sequential view and some simple sequential techniques, this bit of social process would have remained undescribed.

Finding that some transitions are more probable than others

1.8 Hypothesis-generating research

Although we have defined systematic observation in terms of the use of predetermined categories of behavioral codes, we do not think that an investigator need have a coding system before collecting data of interest. For example, Gottman began his research on acquaintanceship in children (1983) by collecting tape recordings of children becoming acquainted. He had no idea at first which situations and experimental arrangements were best for collecting the data. Although the literature suggested some social processes to study, such as communication clarity, conflict resolution, and self-disclosure, he had little idea how to operationalize these constructs. A great deal of work was necessary before a useful coding system was devised. He also found that several different coding systems were necessary to capture different aspects of the children's interaction.

Furthermore, we have found in our own research that as an investigator engages in programmatic research in an area, across a series of studies much is learned about how the initial coding system operates. This leads to revisions of the coding system, and, in some cases, to simplifications. For example, consider conflict resolution in preschool children. An example of a disagreement chain is as follows (Gottman, 1983):

Host (H):	This is stretchy.
Guest (G):	No, it's not.
H:	Uh huh.
G:	Yes.
H:	Uh huh.
G:	It's dirty.
H:	Uh uh.
G:	Uh huh.
H:	Uh uh.
G:	Uh huh.
H:	Uh uh.
G:	Uh huh.
H:	Uh uh. It's not dirty. (p. 27)

These disagreement chains can have fairly neutral affect. An escalated form of conflict involves negative affect (e.g., anger, crying) and is called "squabbling." Preschoolers getting acquainted at home do not squabble very much. Instead, they manage conflict so that it does not escalate. One of the ways that they do this is by giving a reason for disagreeing. This is a sequence in which a child disagrees and then gives a reason. For example: "No, I don't wanna play house. 'Cause I'm not finished coloring." It turns out that this sequence is very powerful in keeping conflict from escalating, compared to interactions in which the disagreement is not as likely to be followed by a reason for disagreeing. This fact was discovered in an initial study on acquaintanceship; this sequence (disagreement is followed by giving a reason for disagreeing) then became an operational definition of a salient social process.

It is perfectly legitimate, in our view, to begin the process of systematic observation with the simple goal of description. As we gain experience with the phenomena we are investigating, we learn which variables are important to us. We can begin our investigation with a search for order. Usually we have some hypotheses about what we might expect to find. The wonderful thing about observational research is that it maximizes the possibility of being surprised.

There is a danger in this hypothesis-generating approach, and this has to do with the temptation of not thinking very much about what one might expect, and instead looking at everything. Our experience in consulting leads us to recognize the danger in this approach. Usually investigators who do not generate hypotheses at all will be overwhe[lme]d by their data. A delicate balance must be worked out that is co[nsistent wit]h the researcher's style.

A basic requirement of this ki[nd of exploratory re]search is that it is essential to replicate and sear[ch for consistency across stu]dies. Our experience with this approach [is that we usually find that] confusing results do not replicate, and th[at interpretable results] do replicate.

To summarize, hy[potheses can and do] play a vital role in the process of descri[bing the flow of the p]henomena. This kind of observational [research is an excellent way] of investigation. However, it needs to be [considered as part of a] n programmatic research that builds in re[plication.]

1.9 Summary: Syst[em]c is not always sequential

We make two assumptions about readers of this book. First, we assume that they are already convinced that systematic observation is an important method for measuring behavior. Second, we assume that they want to explore dynamic, process-oriented aspects of the behavior they observe. This requires that sequential techniques be added to systematic observation.

Systematic observation has two major defining characteristics: the use of predefined behavior codes and a concern with observer reliability. The method is time consuming but offers a degree of certainty and replicability that narrative reports cannot. Even when systematic observation is not sequential, much can be learned, as the work of Parten and of Smith described above demonstrates. In particular, nonsequential systematic observation can be used to answer questions about how individuals distribute their time among various activities, or distribute their utterances among different utterance categories. The data derived from such studies can then be used to ask the usual questions regard-ing how different groups of individuals vary, or how individuals change with age.

Sequential techniques, added to systematic observation, allow a whole new set of questions to be addressed. In particular, sequential techniques can be used to answer questions as to how behavior is sequenced in time, which in turn should help us understand how behavior functions moment to moment. In fact, for purposes of this book, we use the word "sequential" to refer to relatively momentary phenomena, not for developmental phenomena, which are expressed over months or years.

The purpose of this introductory chapter has been to suggest, both in words and by example, what sequential analysis is, why it is useful, and what it can do. In the following chapters, we discuss the various components required of a study invoking sequential analysis.

2

Developing a coding scheme

2.1 Introduction

The first step in observational research is developing a coding scheme. It is a step that deserves a good deal of time and attention. Put simply, the success of observational studies depends on those distinctions that early on become enshrined in the coding scheme. Later on, it will be the job of observers to note when the behaviors defined in the code catalog occur in the stream of behavior. What the investigator is saying, in effect, is: This is what I think important; this is what I want extracted from the passing stream. Yet sometimes the development of coding schemes is approached almost casually, and so we sometimes hear people ask: Do you have a coding scheme I can borrow? This seems to us a little like wearing someone else's underwear. Developing a coding scheme is very much a theoretical act, one that should begin in the privacy of one's own study, and the coding scheme itself represents an hypothesis, even if it is rarely treated as such. After all, it embodies the behaviors and distinctions that the investigator thinks important for exploring the problem at hand. It is, very simply, the lens with which he or she has chosen to view the world.

Now if that lens is thoughtfully constructed and well formed (and aimed in the right direction), a clearer view of the world should emerge. But if not, no amount of corrective action will bring things into focus later. That is, no amount of technical virtuosity, no mathematical geniuses or statistical saviors, can wrest understanding from ill-conceived or wrong-headed coding schemes.

How does one go about constructing a well-formed coding scheme? This may be a little like asking how one formulates a good research question, and although no mechanical prescriptions guaranteeing success are possible, either for coding schemes or for research questions, still some general guidelines may prove helpful. The rest of this chapter discusses various issues that need to be considered when coding schemes are being developed.

2.2 What is the question?

Perhaps the single most important piece of advice for those just beginning to develop a coding scheme is, *begin with a clear question.* For any child, mother–infant pair, couple, animal, or group one might want to study, there are an infinite number of behaviors that could be coded. Without the focus provided by a clear question, it is hard to know just where to look, and it is very easy to be overwhelmed. We have all probably experienced at one time or another how operational goals can take over an activity. For example, a consultant we knew once found the employees of a large college athletic complex diligently recording who used what facility for how long in ways that had become ever more elaborate over the years. It seemed all too easy to think of more complicated ways to encode the passing stream of behavior, but that encoding seemed completely unguided by any purpose. No doubt once there had been some purpose, but apparently it had been forgotten long ago.

Similarly, many investigators, ourselves included, seem tempted to include more and more separate codes, and make ever finer distinctions, simply because it is possible to get observers to record data using such schemes. There is an argument to be made for such a "wide-net" strategy, and it usually goes something like this: Because we are not really sure what will turn out to be important, we need to record everything – or at least lots – and then scan our voluminous data archive for important features. Somehow, it is hoped, coherence will emerge from the multitude. However, we suspect that this happens very rarely, and so when asked, "But what do I do with all this data that I collected so laboriously?" our first suspicion is that a clear question, and not statistical expertise, was lacking. Typically, "categorical overkill" seems to inundate investigators in tedious and not very fruitful detail, whereas studies involving clearly stated questions and tightly focused coding schemes seem far more productive.

For example, consider the following question. Among monkeys, whose early days are spent clinging to their mothers, is it the infants who first leave their mothers as they begin to explore the world, or is it the mothers who first push their reluctant infants out into the world? Given this question, most of us would probably agree on the behavior to be recorded and how it should be categorized. We would want to record separations between mother and infant and categorize each one as being either initiated by the mother or initiated by the infant. Thus our coding scheme would contain two codes – Infant-Initiated Separation and Mother-Initiated Separation – and recording would be "activated" whenever the event of interest – separation – occurred. About the only way to complicate this simple example would be to add a third code: Mutual Disengagement.

With such a simple coding scheme, the progression from data collection to analysis to interpretation would be simple and straightforward. We might find, for example, that over the course of the first several months of life the number of separation episodes gradually increased and then decreased. At first, there would be few separations because the infant is almost always clinging to the mother, whereas later on there might be few because the infant is almost always off the mother, but in between there would be more movement from and to the mother. Further, the data could show that when the number of separations is first increasing, the mother initiated considerably more of them than her infant, whereas later, when the number of separations begins to decrease, the infant initiated more, leading us to conclude that it is the mothers who first push their presumably reluctant infants out into the world.

The point is, developing a coding scheme is theoretical, not mechanical work. In order to work well, a coding scheme has to fit a researcher's ideas and questions. As a result, only rarely can a coding scheme be borrowed from someone else. However, when research questions are clearly stated, it is a much easier matter to determine which distinctions the coding scheme should make. Without clear questions, code development is an unguided affair.

2.3 Physically versus socially based coding schemes

Regardless of whether coding schemes should be borrowed or not – and our previous paragraph objects only to mindless borrowing, not borrowing per se – the fact of the matter is that only rarely are coding schemes used in more than one laboratory. This might seem an undesirable state of affairs to those who believe that scientific work should be replicable. After all, all laboratories interested in temperature use thermometers.

The reason for the commonality of thermometers and the multiplicity of coding schemes has to do, we think, with the nature of the "stuff" being measured – temperature or social behavior. Now it may be that some aspects of social behavior can be measured as precisely and commonly as temperature. Whatever the case, we think it is helpful, when developing a coding scheme, to distinguish between codes that are physically based, on the one hand, and socially based, on the other.

At the risk of precipitating a lengthy discussion with the philosophically inclined, let us state that it seems possible to locate coding schemes along a continuum. One end would be anchored by physically based schemes – schemes that classify behavior with clear and well-understood roots in the organism's physiology – whereas the other end would be anchored by socially based schemes – schemes that deal with behavior whose very classification depends far more on ideas in the mind of the investigator (and others) than on mechanisms in the body. We have called these "socially based," not because they necessarily deal with social behaviors – even though they typically do – but because both their specification and their use depend on social processes. Instead of following quite clearly from physical features or physical mechanisms in a way that causes almost no disagreement, socially based schemes follow from cultural tradition or simply negotiation among people as to a meaningful way to view and categorize the behavior under discussion. Moreover, their use typically requires the observer to make some inference about the individual observed.

For example, some people are paid to determine the sex of baby chicks. The "coding scheme" in this case is simple and obvious: male or female. This is not an easy discrimination to make, and chicken sexers require a fair amount of training, but few people would suggest that the categories exist mainly as ideas in the observers' heads. Their connection with something "seeable," even if difficult to see, is obvious.

Other people (therapists and students influenced by Eric Berne) go about detecting, counting, and giving "strokes" – statements of encouragement or support offered in the course of interaction. In effect, their "coding scheme" categorizes responses made to others as strokes or nonstrokes. For some purposes, therapeutic and otherwise, this may turn out to be a useful construct, but few would argue that "strokes" are a feature of the natural world. Instead, they are a product of the social world and "exist" among those who find the construct useful. Moreover, coding a given behavior as a "stroke" requires making an inference about another's intentions.

Other examples of physically and socially based coding schemes could be drawn from the study of emotion. For example, Ekman and Friesen's (1978) Facial Action Coding System scores facial movement in terms of visible changes in the face brought about by the motion of specific muscle groups (called action units or "AUs"). The muscles that raise the inner corner of the eyebrows receive the code "AU1." The muscles that draw the brows down and together receive the code "AU4." When these muscles act together, they result in a particular configuration of the brow called "AU1+4." The brows wrinkle in specific ways for each of these three action units (see Figure 2.1).

The brow configuration AU1+4 has been called "Darwin's grief muscle" as a result of Darwin's 1873 book on the expression of emotion. This leads us to the point that AU1 and AU1+4 are both brow configurations typically associated with distress and sadness. In fact, there is not always a one-to-one correspondence between sadness or distress and these brow configurations. For example, Woody Allen uses AU1+4 as an underliner

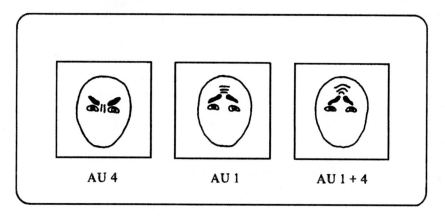

Figure 2.1. Examples of action units from Ekman and Friesen's (1978) Facial Action Coding System.

after he tells a joke. But in most social interaction there are additional cues of sadness or distress.

In a physically based coding scheme we would be coding such things as specific brow configuration, but in a socially based coding scheme we would be coding such things as sadness. The socially based coding scheme requires considerably more inference, and probably requires sensitive observers. However, it is not necessarily less "real" than the system that records brow action. It is simply a different level of description.

Not all researchers would agree with this last statement. Some, especially those trained in animal behavior and ethology, might argue that if the problem is analyzed properly, then any socially based scheme can and should be replaced with a physically based one. We disagree. In fact, we think there are often very good reasons for using socially based coding schemes.

First, it is often the case that physically based schemes, like Ekman and Friesen's mentioned above, are time consuming to learn and to apply, and therefore, as a practical matter, it may be much easier to use a socially based alternative. Even if it is not, a socially based scheme may more faithfully reflect the level of description appropriate for a given research issue. In any given case, of course, investigators' decisions are influenced by the problem at hand and by the audience they want to address, but it seems worth asking, before embarking on an ambitious observational study, whether something simpler, and perhaps more in tune with the research question, would not do as well. Some people will be unhappy with any coding scheme that is not clearly grounded in the physical world, but others, ourselves included, will be tolerant of almost any kind of coding scheme, even one that is quite inferential, as long as it fits the research concerns and as long as trained observers acting independently can agree (see chapter 4).

Second, often various aspects of the socially created world are exactly what we want to study, and in those cases socially based coding schemes are the appropriate "filter" to capture the behavior of interest. Sometimes investigators are advised to avoid coding schemes that require observers to make any judgments about intentionality. In one sense, this is sound advice. Just from the standpoint of observer agreement, there is no reason to ask observers to make more complex judgments (was the baby trying to get her mother's attention?) when simpler ones will do (did the baby approach within 1 meter of her mother and look at her?). But in another sense, if this advice were followed rigorously, we would end up not studying some very interesting behavior, or else defending some possibly quite silly coding schemes.

For example, consider the Parten-like coding scheme for play state, described above, that Peter Smith used. When using this scheme, observers need to discriminate between children playing alone and playing with others, whether in parallel or in a more engaged state. Now one crucial distinction is between being alone and being "with" others; however, whereas observers usually have little trouble agreeing when children are with others, it is not an easy matter to define "withness" in a mechanical way. Is it a matter of proximity? If so, then exactly how near? Is it a matter of visual regard? If so, then how frequently and/or how long? Or does it also include elements of posture, orientation, activity, etc.? We have two choices. Either we define "Alone," "Parallel," and "Group" play in a mechanical way, dependent solely on physical features, and run the risk of occasionally violating our own and our observers' intuitions about the "true" state of affairs; or we view the difference between being alone and being with others as determined by the children's intent and experienced by them subjectively, and we make use of the ability that humans (in this case, our observers) have to judge, more or less accurately (as determined by agreement or consensus), the intentions of others in their culture.

The question is, should we deny or make use of these very human abilities (see Shotter, 1978)? In general, we think we should use them, and that it is especially appropriate to do so when studying social behavior in our fellow humans. (We recognize that this argument begins to break down when animals are being studied.) This is not a new issue in psychological studies, and there is little hope that it will be settled here. Even so, when we argue for more "mentalistic" categories than some traditional psychologists or animal behaviorists would admit, we do not intend this as a license for overly rich and undisciplined speculation and inference. Other things being equal, we much prefer coding categories that are defined in terms of observable and concrete features, but we are not willing to let this one consideration override all others, especially if the meaningfulness of the data collected or the ability of those data to answer the question at hand might suffer. It is possible, for example, to record accurately how many times an infant approaches within 1 meter of his or her mother and what proportion of time was spent in contact with the mother, within 1 meter of the mother, and looking at the mother, and still not be able to gauge validly the quality of the mother–infant relationship. That task might be accomplished better, as Mary Ainsworth, Alan Sroufe, and others have argued, by asking observers to rate, on 7-point scales, how much the infant seeks to maintain contact with the mother, resists contact with the mother, etc., and then assessing the relationship on the basis of the pattern of the rating scales (see Ainsworth, Blehar, Waters, & Wall, 1978; Sroufe & Waters, 1977).

Still, when one is developing coding schemes (or rating scales, for that matter), it is a very useful exercise to describe each behavior (or points on the rating scale) in as specific a way as possible. For example, Bakeman and Brownlee (1980), in their parallel play study, required observers to distinguish between children who were playing alone and who were playing in parallel with others. First, Bakeman and Brownlee viewed videotapes of 3-year-olds in a free play situation. They continually asked each other, is this child in solitary or parallel play? – and other questions, too – and even when there was consensus, they tried to state in as specific terms as possible what cues prompted them to make the judgments they did. They were thus able to make a list of features that distinguished parallel from solitary play; when engaged in "prototypic" parallel play, children glanced at others at least once every 30 seconds, were engaged in similar activities, were within 1 meter of each other, and were no more than 90 degrees away from facing each other directly.

These features were all described in the code catalog (the written description of the coding scheme), but the observers were instructed to treat these defining features as somewhat flexible guidelines and not as rigid mandates. Once they had thoroughly discussed with their observers what most writers mean by parallel play, and had described parallel play in their code catalog, they were willing to let observers decide individual cases on the basis of "family resemblance" to parallel play. We agree with this procedure and would argue as follows: By not insisting that observers slavishly adhere to the letter of the rules, we then make use of, instead of denying, their human inferential abilities. However, those abilities need to be disciplined by discussion, by training, and – perhaps most importantly – by convincing documentation of observer agreement. The result should be more accurate data and data that can "speak" to the complex questions that often arise when social behavior and social development are being studied.

2.4 Detectors versus informants

In the previous section we suggested that it is useful to distinguish between physically based and socially based coding schemes and mentioned a number of implications of this view. We would now like to suggest one final implication of this view, one that affects how the observers' work is regarded and, ultimately, how observers are trained. It is fairly common to regard observers merely as "detectors," as instruments like any other whose job is passively to record what is "there," much as a thermometer records temperature. For physically based coding schemes, this seems like an appropriate view. In such cases, it is not the human ability to make inferences about fellow humans that is being exploited by the investigator, but simply the quite incredible ability of the visual–perceptual system to "see what is there."

For socially based coding schemes, however, it makes sense to regard the observer more as a cultural informant than as a detector. When socially based distinctions are being made, distinctions that typically require some inference about others' intentions, this strikes us as a more accurate way for both the researchers and the observers to think about their task. At least one code catalog that we know of, which was developed during the course of a study of citizen–police interaction, presents the observers' task to them exactly in this way (Wallen & Sykes, 1974). We think that this is a good idea, and that observers using socially based coding schemes do a better job if they appreciate the "participant–observer" aspects of their observation task.

2.5 The fallacy of equating objectivity with physically based schemes

Sometimes physically based coding schemes are mistakenly labeled as more "objective" than socially based systems. Let us consider why this is a fallacy. In the study of affect, one of our colleagues, Mary Lynn Fletcher, pointed out that if you employed observers to study the conversation of Chinese people, it would be absurd to hire observers who could not understand Chinese, claiming that this makes them more "objective." In an analogous way, one can argue that to code complex social processes such as emotions, it is often necessary for observers to be able to understand the "language" of affect. This requires that coders be expert cultural informants about emotion and also be able to respond subjectively. The key to clarifying the issue of "objectivity" is to stress, not objectivity as such, but replicability instead. Independent observers should agree.

There is a second issue inherent in the physically based versus socially based systems, namely, the level of description that is of interest to the investigator. For example, expert cultural informants have a complex catalog of physical cues that they can call upon to decide that a husband is angry in a marital discussion. If an investigator is interested in emotional categories such as angry, sad, contemptuous, etc., then it may be fruitless to attempt to construct a complete list of cues for each emotion judgment. Furthermore, it may not even be possible because the kinds of cues that go into a judgment of anger could be infinite, varying with context, the words spoken, pause, stress, rhythm, amplitude, major or minor key, contours of speech, facial expressions, gestures, and the nonadditive interaction of these features. We may not be interested in a description such as the following:

> At 5 : 19 the husband lowered his brows, compressed his lips, lowered the pitch of his voice, disagreed with his wife, while folding his arms in the akimbo position.

Instead, we may simply wish to describe the husband as angry. As long as we know what we mean by "angry," which will be required if we are ever to achieve reliability between observers, incidents such as the one previously described *illustrate* the construct of anger. They need not *exhaust* it.

In other words, competent observers can make extremely complex social judgments reliably and need not be viewed as inherently undesirable measuring instruments. At times they are exactly what we need, and their subjective observational skills can be precise and the best scientific choice.

2.6 Keeping it simple

Some readers, because of their intellectual convictions, the nature of the problems they study, or both, may not have found the foregoing discussion very helpful. Still, they could probably agree with advice to keep coding schemes simple. Like much good advice, this is so familiar that it seems trite, yet like maintaining liberty, maintaining simplicity also requires constant vigilance. The "natural" press of events seems against simplicity, at least with respect to coding schemes. The investigator's question may be only vaguely apprehended, as discussed earlier, a coding scheme may have been borrowed from someone else inappropriately, the conceptual analysis on which a coding scheme is based may not have been worked through sufficiently, or a proposed scheme may not have been adequately refined through pilot observations and critical discussion. All of these circumstances can confuse the issue and result in overly elaborated and unwieldy coding schemes.

Several points may be useful for keeping the coding system simple. First, it is important to have clear conceptual categories that are essentially at the same level of description. For example, it is not desirable to ask coders to decide whether an utterance was a question and at the same time whether the utterance was responsive to the previous speaker. These judgments are made on two different levels of conceptual complexity. They should not be combined in one coding scheme. Second, codes ought to be reasonably distinct; that is, behavior categories should be homogeneous, which means that even when acts appear somewhat similar, they should not be put in the same category if there is good reason to think that either their causes or their functions are different. Third, it is better to "split" than to "lump" codes. Behavior categories can always be lumped together during data analysis, if that seems useful, but behaviors lumped together by the coding scheme cannot be split out later (unless we have a taped record and recode specific moments).

2.7 Splitting and lumping

At some point in discussions of this sort, the "level of analysis" problem is frequently broached (see, e.g., Hartup, 1979; Suomi, 1979). We can conceive of almost any phenomenon as consisting of levels, hierarchically arranged, with larger and more inclusive or more molar concepts occupying each higher level, and smaller and more detailed or more molecular concepts occupying each lower level. Then the question arises, what level should our coding categories occupy? In the abstract, without the context of a particular question, an absolute answer to this question is hardly possible, but a relative one is. First we need to decide what conceptual level seems appropriate for the question at hand (which is easier said than done); then we should choose coding categories no higher than that level. In fact, there is considerable merit in locating at least some of the categories one level below, on a slightly more detailed level that seems required.

There are at least three reasons why using coding categories that represent a somewhat more molecular level than the level planned for analysis may be a desirable strategy. For example, imagine that our question concerned how often 2-, 3-, and 4-year-old children accompany vocalizations with gestures when directing their attention to others. The question requires only that "bids" for attention be tallied and the presence or absence of vocalizations and of gestures for each bid be recorded. However, we could ask our observers to record whether the bid was directed to an adult or a peer, whether a vocalization involved language or not, whether a nonverbal vocalization was of high or low intensity, whether a verbalization was a question, a command, or a comment, etc. Dropping to a coding level more molecular than required for the question might seem to place additional burdens on the observers, but paradoxically we think that this strategy increases the chances of collecting reliable data. Just as it is often easier to remember three elements, say, instead of one, if those three are structured in some way, so too observers are often more likely to see and record events accurately when those events are broken up into a number of more specific pieces (as long as that number is not too great, of course). This seems to provide the passing stream of behavior with more "hooks" for the observers to grab.

Further, when data are collected at a somewhat more detailed level than required, we are in a position to justify empirically our later lumping. Given the coding scheme presented in the previous paragraph, for example, a critic might object that the different kinds of vocalization we coded are so different that they should not be dealt with as a single class. Yet if we can show that the frequency of gesture use was not different for the different kinds of vocalizations in the different age groups, then there would be no reason, for these purposes, to use anything other than the lumped category. Moreover, the decision to lump would then be based on something more than our initial hunches.

Finally, and this is the third reason, more detailed data may reveal something of interest to others whose concerns may differ from ours, and at the same time may suggest something unanticipated to us. For example, given the coding scheme described above, we might find out that how gestures and vocalizations are coordinated depends on whether the other person involved is a peer or an adult, even if initially we had not been much interested in, nor had even expected, effects associated with the interactive partner.

We should note that often the issue of level of analysis is not the same issue as whether to code data at a detailed or global level. We may have a set of research questions that call for more than one coding system. For example, Gottman and Levenson are currently employing a socially based coding scheme to describe emotional moments as angry, sad, etc. Observers also note if there was facial movement during each emotional moment. These facial movements are then coded with a detailed, physically based coding system, Ekman and Friesen's Facial Action Coding System (FACS). Gottman and Levenson collected psychophysiological data while married couples interacted. One research question concerns whether there are specific physiological profiles for specific categories of facial expressions. The FACS coding is needed to address this question, but in the Gottman and Levenson study a decision had to be made about sampling moments for FACS coding because detailed FACS coding is so costly. The socially based system is thus used as an aid to employing a more detailed, physically based coding system. Coding schemes at different levels of analysis can thus be used in tandem within the same study.

2.8 Mutually exclusive and exhaustive codes

Almost all the examples of coding schemes presented so far consist of mutually exclusive and exhaustive codes. This means that only one code can be associated with a particular event (mutually exclusive) but that there is some code for every event (exhaustive). For example, with respect to Parten's six social participation codes, only one category was appropriate for each 1-minute time sample, but all time samples could be categorized. Observational studies do not require that all coding schemes consist of mutually exclusive and exhaustive codes, but in fact such schemes have several desirable features – their construction requires a certain amount of conceptual analysis, for example, and their use can simplify data analysis – and as a result such schemes are frequently encountered. Some writers even state that coding schemes must consist of mutually exclusive and exhaustive codes, but there are other possibilities, as discussed in the following paragraphs.

In principle, of course, codes for any behavior can be defined in a way that makes them mutually exclusive and exhaustive (ME&E; see S. Altmann, 1965). For example, if we were interested in the coordination of vocal and visual behavior during face-to-face interaction, we might record (a) when person A was looking at, and (b) when person A was talking to his or her partner. Now these two behavioral codes can cooccur and so are not mutually exclusive, but if we regarded their coocurrence as a new or different code, then we could construct an ME&E coding scheme consisting of four codes: (a) A looks, (b) A talks, (c) A both looks and talks, and (d) the "null" code, A neither looks nor talks.

The coding scheme consisting of the two nonmutually exclusive behaviors may offer certain advantages. For one thing, observers have to remember only two behavioral codes and not three (or four, counting the "null" code). When only two nonmutually exclusive "base behaviors" are under consideration, as in the present example, the difference between their number and the number of ME&E "behavior patterns" they generate is not great, but with only a few more base behaviors the number of possible patterns becomes huge. For example, with four base codes there are 16 patterns, with five base codes, 32 patterns, etc., which could rapidly result in observer overload. Moreover, if the times of onsets and offsets for the base behaviors were recorded, data consisting of ME&E categories could always be generated later, if such data were required for subsequent analyses.

It should be noted that coding time may be increased in some ME&E systems because of the nature of the decisions the coder has to make. This applies to schemes that include codes of the sort (a) event A, (b) event B, (c) both event A and B coocurring. If coding time is an issue, an alternative is to have coders use a checklist of items that can cooccur and need not be exhaustive, or a rating system. This is like having coders fill out a brief "questionnaire" after every interactive unit occurs. Coders decide about each "item" of the "questionnaire" independently of every other item. Each item can still be precisely defined.

2.9 The evolution of a coding system

Rosenblum (1978) described the initial stages involved in the creation of a coding system. First, he discussed the importance of establishing the conditions of observation. In particular, the situation selected will affect the diversity of behavior displayed, which will, in turn, determine the complexity of the coding system. Second, he suggested beginning with informal observation of behavior. He wrote,

> [I]t is best to begin in the most unstructured fashion possible. There is great advantage to beginning such observations with only a pencil and blank pad for recording, putting aside the spectres of anthropomorphism, adultomorphism, or any of the other rigidifying constraints that must be imposed in separating wheat from chaff later on in the development of the research program; it is vital to begin by using the incredible synthesizing and integrative functions of the human mind. ... At the beginning, the observer must simply watch, allowing individual subjects to arise as separate entities in the group and behavior patterns to emerge as figures against the background of initially amorphous activity. (pp. 16–17)

We suggest that writing narrative summaries is very helpful at this stage. From the narrative a set of codes is generated, ideally an ME&E set.

As research experience is obtained using a particular coding scheme, it can be modified. For example, Patterson and Moore (1979) discussed using interactive patterns as units of behavior. They analyzed the behavior of Tina using the Family Interaction Coding System (FICS) and found an organization of FICS codes in time. These could later become units of observation. They wrote:

> Examination of Tina's aversive behavior showed that they tended to occur in bursts, followed by periods of neutral or prosocial behavior. ... It seemed, then, that the bursts, themselves, would be a likely area to search next for structure in social interaction. (p. 83)

Next they noted what FICS behavior initiated and maintained these bursts and identified a set of sequences characteristic of the bursts. Using the larger interaction units, they discovered evidence for what they called an "interactional ripple effect," by which they meant the increased likelihood of the initiating event of the chain occurring once a chain has been run off.

There are many consequences of employing a coding system in a series of studies. New codes may appear, new distinctions may be made, or distinctions may disappear as a new lumping scheme is derived. Gottman (1979a) used sequential analysis of a 28-code system for coding marital interaction to create eight summary codes. Two codes were lumped together only if they were functionally similar and sequentially related. For example, the "yes but" code functioned in a similar way to simple disagreement in the sense that both led to disagreement by the spouse, so they were lumped together. Gottman argued against the logic of using factor analysis to lump codes. He wrote:

> The problem with factor analysis in this application is that it lumps two codes on the basis of high correlations between these two codes across subjects. However, just because subjects who do a lot of code A also do a lot of code B does not imply that these two codes are functionally equivalent. (p. 91)

A coding system can evolve as it is being used by intelligent coders. To make this systematic, require observers to write a narrative summary of each observation session, keeping a log of any new behavior that occurs and seems important. Then ask observers to note the contexts in which the new behavior occurs, its antecedents and consequences. The new behavior may be part of a functional behavior set already described, or it may require a category of its own.

2.10 Example 1: Interaction of prehatched chickens

In this chapter and the previous one, we have given several examples of coding schemes. For the most part, these have been quite simple, coding just one kind of behavior, like social participation, with just a few mutually exclusive and exhaustive codes. We would like to end this chapter by describing five somewhat more complex coding schemes. The first involves chickens. In order to study interactions between not-yet-hatched chickens and their setting hens, Tuculescu and Griswold (1983) defined 16 codes, organized as follows:

Embryonic behaviors

Distress-type calls

1. Phioo
2. Soft Peep
3. Peep
4. Screech

Pleasure-type calls

5. Twitter

6. Food Call
7. Huddling Call

Maternal behaviors

Body movements

8. Undetermined Move
9. Egg Turn
10. Resettle

Head movements

11. Peck
12. Beak Clap

Vocalizations

13. Cluck
14. Intermediate Call
15. Food Call
16. Mild Alarm Call

To those familiar with the behavior of chickens, these codes appear "natural" and discrete. Trained observers apparently have no trouble discriminating, for example, between a Phioo, a Soft Peep, a Peep, and a Screech, each of which in fact appears somewhat different on a spectrographic recording. Thus "physical reality" may undergird these codes, but human observers are still asked to make the determinations.

These codes are also clearly organized. There are three levels to this particular hierarchy. On the first level, embryonic and maternal behavior are distinguished; on the second, different kinds of embryonic (distress and pleasure calls) and different kinds of maternal (body movements, head movements, vocalizations) behavior are differentiated; and on the third, the codes themselves are defined. Within each "second-level" category, codes are mutually exclusive, but codes across different second-level categories can cooccur. Indeed, cooccurrence of certain kinds of behavior, like embryonic distress calls and maternal body movements, was very much of interest to the investigators.

There are at least three reasons why we think organizing codes in this hierarchical fashion is often desirable. First, it both ensures and reflects a certain amount of conceptual analysis. Second, it makes the codes easier to explain to others and easier for observers to memorize. Third, it facilitates analysis. For example, for some analyses all embryonic distress calls and all maternal vocalizations were lumped together, which is an example of a practice we recommended earlier – analyzing on a more molar level than that used for data collection.

2.11 Example 2: Children's conversations

The second example is derived from a study conducted by John Gottman (1983) on how children become friends. Working from audiotapes, observers categorized each successive "thought unit" in the stream of talk according to a catalog of 42 mutually exclusive and exhaustive content codes. These 42 codes were grouped into seven superordinate categories: (a) demands for the other child, (b) demands for the pair, (c) you and me statements, (d) self-focus statements, (e) emotive statements, (f) social rules, and (g) information exchange and message clarification statements. Here we reproduce definitions and examples just for the first 16 codes (from Gottman, 1983, p. 13):

Demands for the other child

1. Command (Gimme that.)
2. Polite Requests (That one, please.)
3. Polite Request in Question Form (Would you gimme that?)
4. Suggestion (You could make that black.)
5. Suggestion in Question Form (Why don't you make that black?)
6. Asking Permission (Can I play with that now?)
7. Demands in the Form of an Information Statement (I think my crayons are next to you.)
8. Demands in the Form of a Question for Information (Have you got any sixes?)
9. Wanna (I wanna play house.)
10. Question Wanna (Do you wanna play house?)
11. Requirements for the Other Child (You should stay in the lines.)
12. Asks Help (Would you tie this for me?)

speech acts

We demands (demands for the pair)

13. Hafta Wanna (We have to take a nap.)
14. Let's (Let's play house.)
15. Let's in Question Form (How about drawing now?)
16. Roles to Both (You be the cop and I'll be the robber.)

Unlike with the codes for chicken behavior described above, it is hard to claim that any physical reality undergrids Gottman's content codes. For that very reason, he took considerable pains to demonstrate observer reliability for his codes – which Tuculescu and Griswold did not. Gottman also made finer distinctions when defining his codes than he found useful for later analyses – which is natural tendency when the cleavage between codes is not all that clear. Still, like Tuculescu and Griswold if for slightly different reasons, he found it useful to lump codes for analysis. In fact, the initial 42 content codes were reduced to 20 for his analyses of friendship formation. The three codes derived from the 16 initial codes list above were (a) Weak demands – numbers 2, 3, 5, 6, 7, 8, and 10 above; (b) Strong Demands –

numbers 1, 4, 9, 11, and 12 above; and (c) Demands for the Pair – numbers 13, 14, 15, and 16 above.

Using the result of sequential analysis from a detailed coding system, Gottman devised a "macro" coding system. The macro system was designed so that it would be faster to use (2 hours per hour of tape instead of 30) and would code for the sequences previously identified as important. In the process of building the macro system, new codes were added because in moving to a larger interaction unit, he noticed new phenomena that had never been noticed before. For example, the codes escalation and deescalation of a common-ground activity were created. Gottman (1983) wrote:

> Escalation and deescalation of common-ground activity were included as categories because it appeared that the children often initially established a relatively simple common-ground activity (such as coloring side by side) that made low demands of each child for social responsiveness. For example, in coloring side by side, each child would narrate his or her own activity (e.g., "I'm coloring mine green"). This involved extensive use of the ME codes. Piaget (1930) described this as collective monologue, though such conversation is clearly an acceptable form of dialogue. However, in the present investigation the common-ground activity was usually escalated after a while. This anecdotal observation is consistent with Bakeman and Brownlee's (1980) recent report that parallel play among preschool children is usually the temporal precursor of group play. However, the extent of this process of escalation was far greater than Bakeman and Brownlee (1980) imagined. An example of this escalation is the following: Both children begin narrating their own activity; then one child may introduce INX codes (narration of the other child's activity – e.g., "You're coloring in the lines"); next, a child may begin giving suggestions or other commands to the other child (e.g., "Use blue. That'd be nice"). The activity escalates in each case in terms of the responsiveness demand it places on the children. A joint activity is then suggested and the complexity of this activity will be escalated from time to time.
>
> This escalation process was sometimes smooth, but sometimes it introduced conflict. When it did introduce conflict, the children often deescalated that activity, either returning to a previous activity that they had been able to maintain or moving to information exchange. While many investigators have called attention to individual differences in the complexity of children's dialogue during play (e.g., Garvey, 1974; Garvey & Berndt, 1977), the anecdotal observation here is that a dyad will escalate the complexity of the play (with complexity defined in terms of the responsiveness demand) and manage this complexity as the play proceeds. I had not noticed this complex process until I designed this coding system. However, I do not mean to suggest that these processes are subtle or hard to notice, but only that they have until now been overlooked. An example will help clarify this point. D, the host, is 4-0; and J, the guest, is 4-2. They begin playing in parallel, but note that their dialogue is connected.
>
> 18. J: I got a fruit cutter plate.
> 19. D: Mine's white.

20.   J: You got white Play-Doh and this color and that color.
21.   D: Every color. That's the colors we got.

They continue playing, escalating the responsiveness demand by using strong forms of demands.

29.   D: I'm putting pink in the blue.
30.   J: Mix pink.
31.   D: Pass the blue.
32.   J: I think I'll pass the blue.

They next move toward doing the same thing together (common-ground activity).

35.   D: And you make those for after we get it together, OK? .
36.   J: 'Kay.
37.   D: Have to make these.
38.   J: Pretend like those little roll cookies, too, OK?
39.   D: And make, um, make a, um, pancake, too.
40.   J: Oh rats. This is a little pancake.
41.   D: OK. Make, make me, um, make 2 flat cookies. Cause I'm, I'm cutting any, I'm cutting this. My snake.

The next escalation includes offers.

54.   J: You want all my blue?
55.   D: Yes. To make cookies. Just to make cookies, but we can't mess the cookies all up.
56.   J: Nope.

They then introduce a joint activity and begin using "we" terms in describing what the activity is:

57.   D: Put this the right way, OK? *We're* making supper, huh?
58.   J: *We're* making supper. Maybe *we* could use, if you get white, *we* could use that too, maybe.
59.   D: I don't have any white. Yes, *we*, yes I do.
60.   J: If you got some white, *we* could have some, y'know.

As they continue the play, they employ occasional contextual reminders that this is a joint activity:

72.   D: Oh, we've got to have our dinner. Trying to make some.

D then tries to escalate the play by introducing some fantasy. This escalation is not successful. J is first allocated a low-status role (baby), then a higher-status role (sister), then a higher-status (but still not an equal-status) role (big sister).

76.   D: I'm the mommy.
77.   J: Who am I?
78.   D: Um, the baby.
79.   J: Daddy.
80.   D: Sister.
81.   J: I wanna be the daddy.
82.   D: You're the sister.

83. J: Daddy.
84. D: You're the *big* sister!
85. J: Don't play house. I don't want to play house.

The escalation failure leads to a deescalation.

87. J: Just play eat-eat. We can play eat-eat. We have to play that way.

However, in this case, the successful deescalation was not accomplished without some conflict:

89. J: Look hungry!
90. D: Huh?
91. J: I said look hungry!
92. D: Look hungry? This is dumb.
93. J: Look hungry!
94. D: No!

The children then successfully returned to the previous level of common ground activity, preparing a meal together. Common ground activity is thus viewed in this coding system as a hierarchy in terms of the responsiveness it demands of each child and in terms of the fun it promises. (pp. 55–57)

2.12 Example 3: Baby behavior codes

Our third example describes the behavior of young infants during times when they are being administered a neonatal assessment and was developed by Sharon Landesman-Dwyer (1975). The scheme contains 50 codes grouped into five superordinate categories. The five categories are not mutually exclusive. In fact, they code different possible kinds of behavior, namely, (a) the kind of external stimulation being provided and what the infant is doing with his or her (b) eyes, (c) face (including vocalizations), (d) head, and (e) body. The ten codes within each of these superordinate categories, however, are mutually exclusive and exhaustive. For example, the 10 codes for eyes are:

   0. Can't See
   1. Closed
   2. Slow Roll
   3. Squint
   4. Open–Close
   5. Daze
   6. Bright
   7. Tracking
   8. REM
   9. Other

and the 10 codes for face and vocalization are:

0. Repose
1. Small Move
2. Mouth/Suck
3. Grimace
4. Smile
5. Tremor
6. Yawn
7. Whimper
8. Cry
9. Cry/Tremor

2.13 Example 4: Children's object struggles

The fourth example is taken from a study of the social rules governing object conflicts in toddlers and preschoolers, conducted by Bakeman and Brownlee (1982). They defined six codes, organized into three superordinate categories. Each superordinate category can be cast as a question about an object struggle as follows:

> Prior possession: Did the child now attempting to take an object from another child play with the contested object at any point in the previous minute?
>
> 1. Yes
> 2. No
>
> Resistance: Does the child now holding the object resist the attempted take?
>
> 1. Yes
> 2. No
>
> Success: Is the child who is attempting to take the object successful in getting it?
>
> 1. Yes
> 2. No

The four schemes described above are alike in at least one sense. In each case, the codes defined can be grouped into a relatively small number of superordinate categories. They are also alike in that all codes within a superordinate category are at least mutually exclusive and, with the possible exception of Tuculescu and Griswold, exhaustive as well. Gottman's superordinate categories are themselves mutually exclusive, however, whereas that is not the case for the other three schemes. In fact, the superordinate categories for the other three schemes represent different modes or kinds of behavior or different questions about a particular behavioral event and are clearly not mutually exclusive (with the exception of embryonic distress and pleasure calls).

The Landesman-Dwyer and the Bakeman and Brownlee schemes are formally identical. Both consist of several sets of mutually exclusive and exhaustive codes. This is a useful structure for codes because it ensures that answers to a number of different questions will be answered: What is the baby doing with his eyes? With his mouth? Did the taker have prior possession? They differ, however, in when the questions are asked. Landesman-Dwyer's scheme is used to characterize each successive moment in time, whereas Bakeman and Brownlee's scheme is used to characterize a particular event and is "activated" only when the event of interest occurs.

2.14 Example 5: Monkeys' activities

The fifth example is from the work of Suomi and his coworkers concerning the social development of rhesus monkeys (Suomi, Mineka, & DeLizio, 1983). They defined 14 codes for monkeys' activities that, although not necessarily mutually exclusive, were designed to be exhaustive. Observers were asked to record frequency and duration information for each. The 14 codes (along with brief definitions) were as follows (paraphrased from Suomi et al., 1983, p. 774):

1. Mother–Infant Ventral (mutual ventral and/or nipple contact)
2. Mother–Infant Reject (any successful or unsuccessful break of Mother–Infant Ventral, or rejection of an infant-initiated Mother–Infant Ventral)
3. Ventral Cling (ventral contact with an animal other than the mother)
4. Self-Clasp (clasping any part of own body)
5. Self-Mouth (oral contact with own body)
6. Environmental Exploration (tactual or oral manipulation of inanimate objects)
7. Passive (absence of all social, exploratory, and locomotor activity; could cooccur with self categories and vocalizations)
8. Stereotypy (patterned movements maintained in a rhythmic and repetitive fashion)
9. Locomotion (any self-induced change in location of self, exclusive of stereotypy)
10. Vocalization (any sound emitted by subject)
11. Huddle (a maintained, self-enclosed, fetal-like position)

12. Play and Sex (any type of play and/or sexual posturing, exclusive of locomotion)
13. Aggression (vigorous and/or prolonged biting, hair pulling, clasping, accompanied by one or more of threat, barking, piloerection, or strutting)
14. Social Contact (contact and/or proximity with another subject, exclusive of Mother–Infant Ventral, Ventral Cling, Aggression, or Play and Sex)

Except for Passive (which can cooccur with Self-Clasp, Self-Mouth, and Vocalization), these codes appear to be mutually exclusive. In some cases, activities that could cooccur have been made mutually exclusive by definition. For example, if an activity involves both Stereotypy and Locomotion, then Stereotypy is coded. Similarly, if what appears to be Social Contact involves a more specific activity for which a code is defined (like Play and Sex), then the specific code takes precedence. Defining such rules of precedence is, in fact, a common way to make a set of codes mutually exclusive.

2.15 Summary

No other single element is as important to the success of an observational study as the coding scheme. Yet developing an appropriate scheme (or schemes) is often an arduous task. It should be assumed that it will involve a fair number of hours of informal observation (either "live" or using videotapes), a fair amount of occasionally heated discussion, and several successively refined versions of the coding scheme. Throughout this process, participants should continually ask themselves, exactly what questions do we want to answer, and how will this way of coding behavior help us answer those questions?

There is no reason to expect this process to be easy. After all, quite apart from our current research traditions, developing "coding schemes" (making distinctions, categorizing, developing taxonomies) is an ancient, perhaps even fundamental, intellectual activity. It seems reasonable to view one product of this activity, the coding scheme, as an informal hypothesis, and the entire study in which the coding scheme is embedded as a "test" of that hypothesis. If the "hypothesis" has merit, if it is clearly focused and makes proper distinctions, then sensible and interpretable results should emerge. When results seem confused and inconclusive, however, this state of affairs should not automatically be blamed on a lack of proper data-analytic tools for observational data. First we should ask, are questions clearly stated, and do the coding schemes fit the questions? We hope that a consideration of the various issues raised in this chapter will make affirmative answers to these questions more likely.

In this chapter, we have confined our discussion to coding schemes. Five examples of developed schemes were presented, and an additional four are detailed in the companion volume (Bakeman and Quera, 1995a, chapter 2). We have stressed in particular how coding schemes can be organized or structured and have left for the next chapter a discussion of how coding schemes are put into use. This separation is somewhat artificial. How behavioral sequences are to be recorded can and often does affect how codes are defined and organized in the first place. This is especially evident when the Landesman-Dwyer and the Bakeman and Brownlee schemes discussed earlier are compared. Still, a scheme like Landesman-Dwyer's could be recorded in two quite different ways, as we discuss in the next chapter. It is the task of that chapter to describe the different ways behavior can be recorded, once behavioral codes have been defined.

3

Recording behavioral sequences

3.1 Recording units: Events versus intervals

The title of this chapter, Recording behavioral sequences, suggests two different but somewhat related topics. The chapter could deal primarily with mechanical matters and describe devices used to record data, or it could deal just with conceptual matters and describe different strategies for collecting data about behavior sequences. Actually, this chapter will be a mixture of both, although conceptual and strategic matters will be stressed.

One important issue has already been raised by the example of the Bakeman and Brownlee (1980) study of parallel play in chapter 1, and that is the issue of "units." Before selecting a particular recording strategy, an investigator needs first to decide what "units" are to be used for recording.

As we use the term, the recording unit identifies what prompts the observer to record, and usually is either an interval or an event. For example, an investigator might choose to code time intervals, as Bakeman and Brownlee did, assigning codes to successive time intervals. This is a common strategy, but for many purposes more accurate data result when the events themselves are coded instead. In such cases, observers wait for an event of interest to occur. When one occurs, they code it (i.e., note what kind of event it was) and perhaps record onset and offset times for the event as well.

Which is better, to code events or to code intervals? That depends on a number of factors, including the kind and complexity of the coding scheme, the desired accuracy for the data, and the kind of recording equipment available. These issues are discussed later on in this chapter in the context of talking about applications of specific recording strategies, but first we want to make another distinction, one between momentary and duration events.

3.2 Momentary versus duration events

Sometimes investigators are concerned only with how often certain events occur, or in what order they occur, and are not much concerned with how long they last. At other times, duration – the mean amount of time a particular kind of event lasts or the proportion of time devoted to a particular kind of event – is very much of concern. As a result, many writers have found it convenient to distinguish between "momentary events" (or frequency behaviors) on the one hand, and "behavioral states" (or duration behaviors) on the other (J. Altmann, 1974; Sackett, 1978). The distinction is not absolute, of course, but examples of relatively brief and discrete, momentary events could include baby burps, dog yelps, child points, or any of Gottman's thought unit codes described in section 2.11, whereas examples of duration events could include baby asleep, dog hunting, child engaged in parallel play, or any of Landesman-Dwyer's baby behavior codes described in section 2.12.

One particular way of conceptualizing duration events is both so common and so useful it deserves comment. Often researchers view the events they code as "behavioral states." Typically, the assumption is that the behavior observers see reflects some underlying "organization," and that at any given time the infant, animal, dyad, etc., will be "in" a particular state. The observers' task then is to segment the stream of behavior into mutually exclusive and exhaustive behavioral states, such as the arousal states often described for young infants (REM sleep, quiet alert, fussy, etc.; Wolff, 1966).

The distinction between momentary and duration events (or between discrete events and behavioral states) seems worth making to us, partly because of the implications it may have for how data are recorded. When the investigator wants to know only the order of events (for example, Gottman's study of friendship formation) or how behavioral states are sequenced (Bakeman and Brownlee's study of parallel play), then the recording system need not preserve time. However, if the investigator wants also to report proportion of time devoted to the different behavioral states, then time information of course needs to be recorded. In general, when duration matters, the recording system must somehow preserve elapsed time for each of the coded events. Moreover, when occurrences of different kinds of events are to be related, beginning times for these events need to be preserved as well. (Examples are provided by Landesman-Dwyer and by Tuculescu and Griswold; see section 3.5, Recording onset and offset times.)

3.3 Continuous versus intermittent recording

Before discussing particular recording strategies as such, there is one more distinction that we would like to make. Almost all strategies we describe here are examples of continuous, not intermittent, recording. The phrase "continuous recording" evokes an event recorder, with its continuously

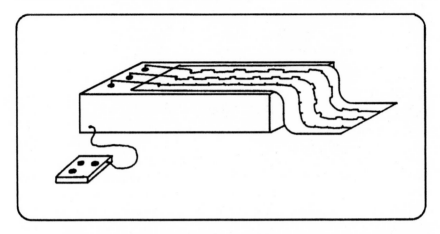

Figure 3.1. An example of continuous recording: A deflected-pen event recorder.

moving roll of paper and its pens, ready to record events by their deflection (see Figure 3.1). It is a rather cumbersome device, rarely used in observational research. Almost always, researchers prefer pencil, paper, and some sort of clock, or else (increasingly) an electronic recording device. Still, the phrase "continuous recording" seems appropriate for the strategies we describe here, not because the paper rolls on, but because the observers are continuously alert, paying attention, ready to record whenever an event of interest occurs, whenever a behavioral state changes, or whenever a specific time interval elapses.

Given that this is a book about sequential analysis in particular, and not just systematic observation in general, the emphasis on continuous recording is understandable. After all, for sequential analysis to make much sense, the record of the passing stream of behavior captured by the coding/recording system needs to be essentially continuous, free of gaps. However, we do discuss intermittent recording in section 3.9 (Nonsequential considerations: time sampling).

The purpose of the following sections is to describe different ways of collecting observational data, including recording strategies that code events and ones that code intervals. For each way of collecting data, we note what sort of time information is preserved, as well as other advantages and disadvantages.

3.4 Coding events

In this section we shall discuss event-based coding. The basic aspect to take note of is the observer's task, and, in particular what gets the observer

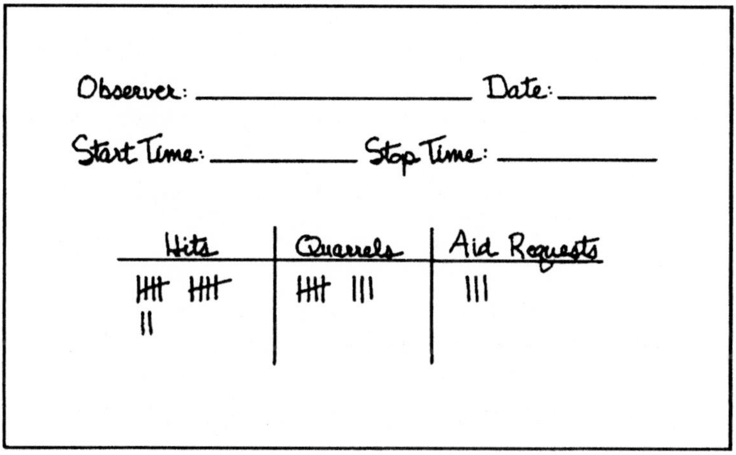
Observer:
Date:
Start Time:
Stop Time:
Hits
Quarrels
Aid Requests

Figure 3.2. Tallying events with a simple checklist.

to record a particular code. When the events of interest, not a time interval running out, are what stir an observer into action, we would say that an event coding strategy is being used to record observational data. The simplest example of event coding occurs when observers are asked just to code events, making no note of time. For example, an investigator might be interested in how often preschool children try to hit each other, how often they quarrel, and how often they ask for an adult's aid. The observer's task then is simply to make a tally whenever one of these codable events occurs. Such data are often collected with a "checklist." The behavioral codes are written across a page, at the top of columns. Then when a codable event occurs, a tally mark is made in the appropriate column. No doubt our readers are already quite aware of this simple way of collecting data. Still, it is useful whenever investigators want only to know how often events of interest occur (frequency information) or at what rate they occur (relative frequency information) (see Figure 3.2).

Such data can be important. However, of more immediate concern to us, given the focus of this book, are event coding strategies that result in sequential data. For example, Gottman segmented the stream of talk into successive thought units. Each of these events was then coded, providing a continuous record of how different kinds of thought units were sequenced in the conversations Gottman tape-recorded. Similarly, Bakeman and Brownlee (who actually used an interval coding strategy) could have asked observers to note instead whenever the play state of the child they were observing changed. Each new play state would then have been coded, resulting in a record of how different kinds of play states were sequenced during free play (see Figure 3.3).

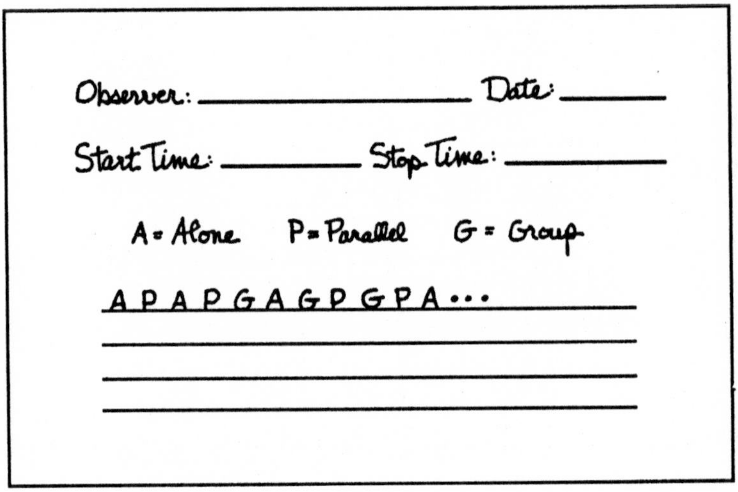

Figure 3.3. Recording the sequence of events.

In these two examples, the basic requirement for sequential data – continuity between successive coded units – is assured because the stream of talk or the stream of behavior is segmented into successive events (units) in a way that leaves no gaps. However, sequential data may also result when observers simply report that this happened, then that happened, then that happened next, recording the order of codable events. Whether such data are regarded as sequential or not depends on how plausible the assumption of continuity between successive events is, which in turn depends on the coding scheme and the local circumstances surrounding the observation. However, rather than become involved in questions of plausibility, we think it better if codes are defined so as to be mutually exclusive and exhaustive in the first place. Then it is easy to argue that the data consist of a continuous record of successive events or behavioral states.

Whether behavior is observed "live" or viewed on videotape does not matter. For example, observers could be instructed to sit in front of a cage from 10 to 11 a.m. on Monday, 3 to 4 p.m. on Tuesday, etc., and record whenever an infant monkey changed his activity, or observers could be instructed to watch several segments of videotape and to record whenever the "play state" of the "focal child" changed, perhaps using Smith's social participation coding scheme (Alone, Parallel, Group). In both cases, observers would record the number and sequence of codable events. An obvious advantage of working from videotapes is that events can be played and replayed until observers feel sure about how to code a particular sequence.

Still, both result in a complete record of the codable events that occurred during some specified time period.

3.5 Recording onset and offset times

When time information is required, which usually means that the investigator wants to report time-budget information or else wants to report how different kinds of behavior are coordinated in time, observers can be asked to record, not just that a codable event occurred, but its onset and offset times as well. This is one way to preserve time information, but it is not the only way, as we discuss in the next section.

The task is made even easier when codes are mutually exclusive and exhaustive (or consist of sets of ME&E codes) because then offset times do not need to be recorded. In such cases, the offset of a code is implied by the onset of another mutually exhaustive code. As an example, consider Landesman-Dwyer's codes for baby's eyes (Closed, Slow Roll, Daze, Bright, etc.). She could have instructed observers to record the time whenever the "state" of the baby's eyes changed – for example, from Closed to Slow Roll. Then the elapsed time or duration for this episode of Closed would be the onset time of Slow Roll minus the onset time of Closed. Moreover, because the times when the baby's eyes were closed is known, she could also ask what kinds of face, head, and body movements occurred when the baby's eyes were closed.

This strategy of recording onset times (or onset and offset times when codes are not mutually exclusive and exhaustive) seems so simple, general, and straightforward to us that we are surprised that it is not used more often. The reason for this, we believe, is primarily technical. Before electronic devices that automatically record time became available, we suspect that observers found it distracting to write down times. Thus the older literature especially contains many examples of a "lined paper" approach to recording time information. Behavioral codes would head columns across the top of the page. Then each row (the space between lines) would represent a period of time. Observers would then make a check or draw a vertical line in the appropriate column to indicate when that kind of event began and how long it lasted. As we discuss later, such an "interval coding" strategy has the merit of requiring only pencil and paper, but it remains an approximation of what researchers often really want to do, which is to record the exact times when the behavior of interest begins and ends.

Timing onsets and offsets does not require electronic recording devices, of course, but such devices do make the task easy. In this paragraph and the next, we describe two applications, one that uses such devices and one

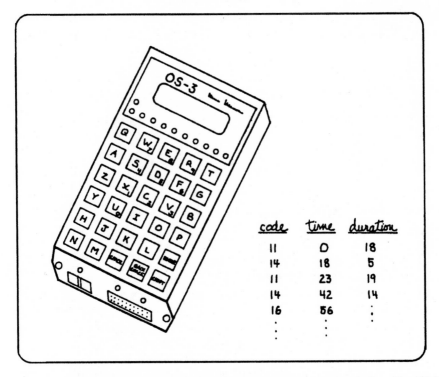

Figure 3.4. An example of a hand-held electronic recording device: The OS-3 (Observational Systems, Redmond, WA 98052). Data production in the form shown is just one of many options.

that does not. First, imagine that observers using the Landesman-Dwyer baby's eyes code were equipped with electronic recording devices. They would learn to push the correct keys by touch; thus they could observe the baby continuously, entering codes to indicate when behavior changed. (For example, an 11 might be used for Closed, a 12 for Slow Roll, a 16 for Bright, etc.) Times would be recorded automatically by a clock in the device (see Figure 3.4). Later, the observer's record would be "dumped" to a computer, and time budget (i.e., percentage scores for different behavioral codes) and other information would be computed by a computer program. (In fact, Landesman-Dwyer used such devices but a different recording strategy, as described in the next section.)

The second application we want to describe involves pencil and paper recording, but couples that with "time-stamped" videotapes. For a study of preverbal communication development, Bakeman and Adamson (1984) videotaped infants playing with mothers and with peers at different ages.

At the same time that the picture and sound were recorded, a time code was also placed on the tape. Later, observers were instructed to segment the tapes into different "engagement states" as defined by the investigators' code catalog. In practice, observers would play and replay the tapes until they felt certain about the point where the engagement state changed. They would then record the code and onset time for the new engagement state.

In summary, one way to preserve a complete record of how behavior unfolds in time (when such is desired) is to record onset (and if necessary offset) times for all codable events. This is easy to do when one is working from time-stamped videotapes. When coding live, this is probably best done with electronic recording devices (either special purpose devices like the one shown in Figure 3.4 or general-purpose handheld or notebook computers, programmed appropriately, which increasingly are replacing special purpose devices). When such devices are not available, the same information can be obtained with pencil, paper, and some sort of clock, but this complicates the observers' task. In such cases, investigators might want to consider the approximate methods described in section 3.7, on Coding intervals.

3.6 Timing pattern changes

There is a second way of preserving complete time information, but it applies only to coding schemes structured like the Landesman-Dwyer Baby Behavior Code described earlier. That is, the scheme must consist of groups of mutually exclusive and exhaustive codes, with each group referring to a different aspect of behavior. Paradoxically, this approach seems like more work for the observers, but many observers (and investigators) prefer it to the simple recording of onset times discussed above.

To describe this recording strategy, let us continue with the example of the Baby Behavior Code. Recall that there were five groups of codes: External Stimulation, Eyes, Face, Head, and Body. Observers are taught to think of this as a 5-digit code: The first digit represents the kind of External Stimulation, the second digit represents the Eyes, the third the Face, etc. Thus the code 18440 means that external stimulation was a reflex (code 1), REM movement was evident in the eyes (code 8), there was a smile on the face (code 4), the head was up (also code 4), and the body was in repose (code 0). If code 18440 were followed by code 18040, it would mean that nothing had changed with respect to External Stimulation, Eyes, Head, and Body but that the Face was now in repose (code 0), not smiling as before.

Each time there is a change in codable behavior, even if it involves only one of the five superordinate groups, a complete 5-digit code is entered. On the surface of it, this seems like more work than necessary. After all, why require observers to enter codes for external stimulation, eyes, head, and body when there has been no change just because there has been a change in facial behavior? In fact, observers who use this approach to recording data report that, once they are trained, it does not seem like extra work at all. Moreover, because the status of all five groups is noted whenever any change occurs, investigators feel confident that changes are seldom missed, which they might not be if observers were responsible for monitoring five different kinds of behavior separately.

Paradoxically, then, more may sometimes be less, meaning that more structure – that is, always entering a structured 5-digit code – may seem like less work, less to remember. This is the approach that Landesman-Dwyer actually uses for her Baby Behavior Code. We should add, however, that her observers use an electronic recording device so that time is automatically recorded everytime a 5-digit code is entered. In fact, we suspect that the Baby Behavior Code would be next to impossible to use without such a device, no matter which recording strategy (timing onsets or timing pattern changes) were employed. Given proper instrumentation, however, the same information (frequencies, mean duration, percents, cooccurrences, etc.) would be available from data recorded using either of these strategies. When a researcher's coding scheme is structured appropriately, then, which strategy should be used? It probably depends in part on observer preference and investigator taste, but we think that recording the timing of pattern changes is a strategy worth considering.

3.7 Coding intervals

For the recording strategies just described – coding events, recording onset and offset times, recording pattern change times – observers are "event triggered," that is, they are stirred to record data whenever a codable event occurs. When using an interval coding (or interval recording) strategy, on the other hand, observers are "time triggered," that is, typically they record at certain predetermined times.

The essence of this strategy is as follows. A child, a dyad, an animal, a group, or whatever, is observed for a period of time. That period is divided into a number of relatively brief intervals, typically on the order of 10 or 15 seconds or so. Observers then either categorize each interval or else note which codable events, if any, occurred during each interval.

Such data can be collected easily with pencil and paper via a "checklist" format. As mentioned briefly earlier, behavioral codes head columns across the top of a page, whereas the rows down the page represent successive

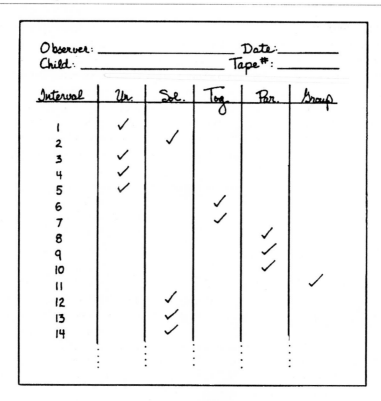

Figure 3.5. Coding intervals using a checklist format.

intervals of time. When a codable event happens within an interval, observers place a check in the appropriate row and column. Even though observers may actually first move pencil to paper when an event happens, and so are "event-triggered" in that sense, the end result is to characterize intervals as containing or not containing an event of interest, which is why we call this an "interval coding" strategy.

An even simpler example of interval coding is provided by the Bakeman and Browenlee study of parallel play, described in chapter 1. Their observers watched 12–15 7-minute video recordings for each of 41 children. The tapes were played at normal speed without stopping, in effect simulating live "real time" observation. Every 15 seconds, a simple electronic device delivered a "click" to the ear, at which time observers wrote down the social participation category that they felt best characterized the time since the last click (see Figure 3.5).

Interval coding, or some variant of it, has been both widely used and much criticized in the literature. We think the reason it has been so widely used has something to do with historical precedent, a human propensity to impose clock units on passing time, the ready availability of lined paper, and the unavailability until quite recently of electronic recording devices. It has been criticized because it simply cannot provide as accurate information as an event coding strategy. The major problem is that more than one code may occur within an interval or near the boundaries of intervals. What would a coder do in this case? If a hierarchical decision rule is employed, sequential and other information within the interval is lost. This problem is minimized, of course, if the time intervals are short relative to the behavior observed.

However, we can see no good theoretical reason ever to use interval coding; its merits are all practical. It requires only pencil, paper, and some sort of simple timing device. No sophisticated electronic recording devices or computers need be involved. In addition, as mentioned earlier, sometimes observers find it easier to categorize intervals than to identify when codable events began. (We suspect, however, that this is more of a consideration when behavior is recorded live, and is much less of a consideration when observers can play and replay videotaped behavior, as for the Bakeman and Adamson study of infant engagement states described in section 3.5.)

Clearly, the key consideration when an interval coding strategy is used is the length chosen for the interval. If that interval is somewhat shorter than the shortest duration typically encountered for a codable event, then little distortion should be introduced into the data (Smith & Connolly, 1972; cf. J. Altmann, 1974). No doubt, most investigators who use an interval coding strategy understand perfectly well that the interval used should not be so long as to mask onsets and offsets of the events being studied because if it is, then not only will estimates of frequencies, durations, and percentages be inaccurate, but behavioral sequences will be distorted as well.

For example, imagine that the codable event is baby within arm's length of mother and that the interval used for recording is 10 seconds. We should be able to assume that when a checked interval follows an unchecked interval, the baby approached the mother. More importantly, we should also be able to assume that when a string of checked intervals follow each other, the baby did not leave the mother and return but rather stayed near her. Clearly, an interval of 5 minutes would not be satisfactory, 10-second intervals are probably acceptable, and 1-second intervals might be even better. Whether the 1-second interval provides sufficiently more accurate data to be worth the extra effort, however, is debatable, although the matter could be determined with a small pilot study.

In sum, when simplicity and low cost of instrumentation matter more than accuracy, or when the interval is shorter than most codable events and observers prefer checking or categorizing intervals to recording onset times, then an interval coding strategy probably makes sense. For all other cases, some variant of event coding should probably be used.

3.8 Cross-classifying events

In sections 3.4, 3.5, and 3.6, we referred to event coding. By that we meant a recording strategy that requires observers to detect codable events whenever they occur in the passing stream of behavior. Once a codable event has been detected, it is then the observers' task to classify the event, that is, to assign to it one of the codes from the coding scheme. This strategy becomes a way of recording behavioral sequences when continuity between successive events can be assumed, as was discussed in section 3.4.

There is a second way of coding events, however, which does not require any kind of continuity between successive events but which nonetheless results in behavioral sequences being captured. This method does not simply classify events (on a single dimension) but instead cross-classifies them (on several dimensions). The key feature of this approach is the coding scheme. Sequential data result when the superordinate categories of the scheme represent logically sequential aspects of the event.

For example, imagine that observers were asked to note whenever children quarreled. If observers did only this, the result would be a frequency count for quarrels but no sequential information. However, observers could also be asked to note what the children were doing just before the quarrel began, what kind of quarrel it was, and how the quarrel was resolved. Assuming that a mutually exclusive and exhaustive set of codes was defined for each of these three questions, the observer would be cross-classifying the quarrel. This is essentially the same kind of task as asking a child to classify a set of objects by shape (circles, squares, and triangles), color (red, blue, and green), and material (wood, metal, plastic). However, there is a key difference between these two classification tasks. The three schemes used to cross-classify quarrels have a natural temporal order – preceding circumstance, quarrel, resolution – whereas the three schemes used to cross-classify objects do not. Thus the strategy of cross-classifying events is not necessarily sequential. Whether it is or not depends on how the coding scheme is defined.

A second example of this strategy is provided by the Bakeman and Brownlee study of object struggles. (Their coding scheme was described in section 2.13.) In that study, the event of interest was an object struggle. Whenever one occurred, observers recorded (a) whether the child attempting to take the object had had prior possession or not, (b) whether the current possessor resisted the take attempt or not, and (c) whether the child attempting to take the object was successful or not. Note that the possible answers to each of these three questions (yes/no) are mutually exclusive and exhaustive and that the three questions have a natural temporal order.

The cross-classification of events is a recording strategy with many advantages. For one thing, techniques for analyzing cross-classified data (contingency tables) have received a good deal of attention, both historically and currently, and they are relatively well worked out (see chapter 10). Also, clear and simple descriptive data typically result. For example, in another study, Brownlee and Bakeman (1981) were interested in what "hitting" might mean to very young children. They defined three kinds of hits (Open, Hard, and Novelty) and then asked observers to record whenever one occurred and to note the consequence (classified as No Further Interaction, Ensuing Negative Interaction, or Ensuing Positive Interaction). They were able to report that open hits were followed by no further interaction and novelty hits by ensuing positive interaction more often than chance would suggest, but only for one of the age groups studied, whereas hard hits were associated with negative consequences in all age groups.

A further advantage is that a coding scheme appropriate for cross-classifying events (temporally ordered superordinate categories, mutually exclusive and exhaustive codes within each superordinate category) implies a certain amount of conceptual analysis and forethought. In general, we think that this is desirable, but in some circumstances it could be a liability. When cross-classifying events, observers do impose a certain amount of structure on the passing stream of behavior, which could mean that interesting sequences not accounted for in the coding scheme might pass by unseen, like ships in the night. Certainly, cross-classifying events is a useful and powerful way of recording data about behavioral sequences when investigators have fairly specific questions in mind. It may be a less useful strategy for more exploratory work.

3.9 Nonsequential considerations: Time sampling

The emphasis of this book is on the sequential analysis of data derived from systematic observation. However, not all such data are appropriate for sequential analyses. Recognizing that, we have stressed in this chapter recording strategies that can yield sequential data. There exist other useful and widely used recording strategies, however, which are worth mentioning if only to distinguish them from the sequential strategies described here.

Perhaps the most widely used nonsequential approach to recording observational data is time sampling, or some variant of it. We have already discussed this approach in reference to Parten's classic study of social participation described in chapter 1. The essence of time sampling is that observing is intermittent, not continuous. Repeated noncontiguous brief periods of time are sampled, and something about them is recorded. There is no continuity between the separate samples, and thus the resulting data are usually not appropriate candidates for sequential analysis.

A variant of time sampling, which at first glance might appear sequential, requires observers to intersperse periods of observing and periods of recording. For example, an observer might watch a baby orangutan for 15 seconds, then record data of some sort for the next 15 seconds, then return to a 15-second observation period, etc. Assuming that the behavior of interest occurs essentially on a second-by-second basis, this might be a reasonable time-sampling strategy, but it would not produce sequential data. Even if the events being coded typically lasted longer than 15 seconds, it would still be a dubious way to collect sequential data because of all the gaps between observation periods.

For every general rule we might state, there are almost always exceptions. Imagine, for example, that we were interested in how adults change position when sleeping. One obvious way to collect data would be to use time-lapse photography, snapping a picture of the sleeping person every minute or so. Now surely this fits the definition of a time-sampling strategy. Yet assuming that sleeping adults rarely change position more than once within the time period that separates samples, such data would be appropriate for sequential analysis. Logically, such data are not at all different from the data produced for Bakeman and Brownlee's study of parallel play described in chapter 1. In each case, the data consist of a string of numbers, with each number representing a particular code. The only difference lies with the unit coded. For the sleeping study, the unit would be an essentially instantaneous point in time, whereas for the parallel play study, observers were asked to characterize an entire period of time, integrating what they saw.

As the previous paragraph makes clear, time sampling can be regarded as a somewhat degenerate form of interval coding (section 3.7 above). Nonetheless, sequential data may result. Whether it does or not, however, is a conceptual matter, one that cannot be mechanically decided. To restate, whether data are regarded as sequential depends on whether it is reasonable to assume some sort of continuity between adjacent coded units. Usually, but not always, time-sampled data fail this test. This does not mean that a time-sampling recording strategy is of little use, only that it is often not appropriate when an investigator's concerns are primarily sequential. However, when researchers want to know how individuals spend time, time sampling may be the most efficient recording strategy available.

3.10 The pleasures of pencil and paper

Throughout this chapter, we have made occasional comments about the mechanics of data recording. Although there are other possibilities, mainly we have mentioned just two: pencil and paper methods, on the one hand, and the use of electronic recording devices, on the other. In this section, we would like to argue that pencil and paper have their pleasures and should not be shunned simply because they seem unsophisticated.

Pencil and paper methods have many advantages. For one thing, it is difficult to imagine recording instruments that cost less or are easier to replace. There are no batteries that may at a critical moment appear mysteriously discharged. Also, pencil and paper are easy to transport and use almost anywhere. Moreover, there is a satisfying physicality about pencil marks on paper. The whole record can be seen easily and parts of it modified with nothing more complicated than an eraser. Almost never does the record of an entire observation session disappear while still apparently is one's hands. Although paper can be lost, it almost never malfunctions.

Pencil and paper methods are equally usable when the researcher is observing live behavior, is viewing videotapes, or is working from a corpus of written transcripts. For example, pencil and paper recording was used for Brownlee and Bakeman's study of hitting, mentioned in section 3.8 (children were observed live for the study proper, although researchers developed the codes while watching videotapes); for Bakeman and Adamson's study of communication development in infants, referred to in section 3.5 (observers worked with videotapes, stopping, reversing, and replaying as appropriate); and for Gottman's study of friendship formation, described in section 2.11 (observers worked with transcripts, segmented into thought units).

3.11 Why use electronics?

As a general rule, users are attracted to more complex and sophisticated devices of any kind because, once mastered, savings in time and labor result. The key phrase here is "once mastered," because almost inevitably, more sophisticated devices – such as electronic recording devices instead of pencil and paper – require that the user pay an "entry cost." This entry cost includes, not just the initial cost of the device, but the time it takes to learn how to use the device and to keep the device working properly. Nonetheless, many researchers find that the entry costs of electronic recording devices are well worth paying. (To our knowledge, such devices have no common generic name or acronym; they are not, for example commonly called ERDs.)

Electronic recording devices can be either special purpose (see Figure 3.4) or, increasingly, general purpose laptop or notebook computers, programmed to collect data for a particular study. A major advantage of such devices is that data are recorded in machine-readable form from the start. There is no need for data to pass through the hands of a keypuncher, and hence keypunchers cannot introduce errors into data later. Such devices work as follows. To indicate the sequence of codable events, observers depress keys corresponding to the appropriate behavioral codes. A record of the keys "stroked" is stored in an internal memory. Once the observation session is complete, the contents of memory can be transferred to a file on the computer's hard disk or written to some external storage medium, like a floppy disk or a magnetic backup tape. The stored data are then available for whatever subsequent processing is desired.

A second major advantage of electronic recording devices is that they usually contain an internal clock. This means that whenever an observer enters a code, the time can be stored as well, automatically, without the observer having to read the time. Reading the time and writing it down are, of course, quite distracting for an observer. That is why we think interval coding has been used so often in the past (a click every 10 seconds, for example, can be delivered to the ear, leaving the eye free) and why, given the availability of electronic recording devices, event coding (with the automatic recording of onset and offset times or of pattern changes) is becoming much more common than in the past.

A final advantage likewise has to do with minimizing distractions, leaving observers free to devote their full attention to whatever or whomever is being observed. Observers easily learn to push the appropriate keys without looking at them, like a good typist. With these keyboard devices, then, observers do not need to shift their eyes away from the recording task, as they do when recording with pencil and paper. This matters most, of course, when observing live.

Electronic devices for recording data, then, are recommended when a machine-readable record is required, and it seems advantageous to avoid a keypunching step during data preparation. They are also recommended when observers need to keep their eye on the task without the distractions of writing down codes, or worse, their times of occurrence. A code such as the Baby Behavior Code described in section 3.6, which requires the frequent entering of 5-digit codes along with the times they occurred, is probably unworkable with anything other than an electronic recording device.

Other, more sophisticated possibilities for recording data electronically should be mentioned. One common strategy involves time-stamped videotapes (i.e., videotapes on which the time, usually accurate to the nearest second or some fraction of a second, forms part of the picture). Observers

Table 3.1. *Summary of recording schemes*

Recording scheme	Definition	Advantages	Disadvantages
Event recording	Events activate coders	May provide a realistic way of segmenting behaviors	Could lose time information unless onsets and offsets were noted
Interval recording	Time activates coders	Easy to use	1. May artificially truncate behavior 2. Need to select interval small enough or could lose information
Cross-classifying events	Events activate coders, but only a specific kind of event	1. Speed 2. Can still preserve sequential information 3. Statistics well worked out 4. Requires conceptual forethought	Could lose valuable information not accounted for by the highly selective scheme
Time sampling	Time activates coders	Easy to use	Coding is intermittent so sequential information is usually lost

view tapes, slowing down the speed and rewinding and reviewing as necessary. The times when events occurred are read from the time displayed on the screen. This time may be written down, using pencil and paper, or keyed along with its accompanying code directly into a computer. An improvement on this strategy involves recording time information on the videotape in some machine-readable format and connecting a computer to the video player. Then observers need only depress keys corresponding to particular behaviors; the computer both reads the current time and stores it along with the appropriate code. Moreover, the video player can be controlled directly from the computer keyboard. Then, with appropriate hardware and software, coders can instruct the system to display all segments of tape previously coded for a particular code or set of codes. Such systems are very useful but, as you might expect, have a relatively high entry cost both in terms of money and time (e.g., see Tapp & Walden, 1993).

3.12 Summary

Table 3.1 is a summary of the four major conceptual recording schemes we have discussed in this chapter, together with their advantages and disadvantages. The particular recording scheme chosen clearly depends on the research question. However, in general, we find event recording (with or without timing of onsets and offsets or timing of pattern changes) and cross-classifying events to be more useful for sequential analyses than either interval recording or time sampling.

4

Assessing observer agreement

4.1 Why bother?

Imagine that we want to study how objects are used in communication between mothers and their 15-month-old infants, and have made a number of videotapes of mothers and infants playing together. We might focus our attention on those times when either the mother or the infant tries to engage the other's interest in some object, and then describe what those attempts look like and what their consequences are. After viewing the videotapes several times, sometimes in slow motion, we might be convinced that we had detected all such episodes, had accurately described them, and were now ready to make statements about how mothers and infants go about attempting to interest each other in objects.

We should not be surprised, however, if other investigators do not take our conclusions as seriously as we do. After all, we have done nothing to convince them that others viewing the videotapes would see the same things, much less come to the same conclusions. We are probably all aware how easy it is to see what we want to see, even given the best of intentions. For that reason, we take elaborate precautions in scientific work to insulate measuring procedures from the investigator's influence.

When measurements are recorded automatically and/or when there is little ambiguity about the measurement (for example, the amount of sediment in a standard sample of seawater), as is often the case in the "hard" sciences, the problem of investigator bias is not so severe. But in observational studies, especially when what we call "socially based" coding schemes are being used, it becomes especially important to convince others that what was observed does not unduly reflect either the investigator's desires or some idiosyncratic worldview of the observer. The solution to the first problem is to keep observers naive as to the hypotheses under investigation. This in fact is done by most investigators, judging from the reports they write, and should be regarded as standard practice in observational work. The solution to the second problem is to use more than one observer and to assess how well they agree.

Accuracy

The major conceptual reason for assessing interobserver agreement, then, is to convince others as to the "accuracy" of the recorded data. The assumption is, if two naive observers independently make essentially similar codings of the same events, then the data they collect should reflect something more than a desire to please the "boss" by seeing what the boss wants, and something more than one individual's unique and perhaps strange way of seeing the world.

Some small-scale studies may require only one observer, but this does not obviate the need for demonstrating agreement. For example, in one study Brownlee and Bakeman (1981) were concerned with communicative aspects of hitting in 1-, 2-, and 3-year-old children. After repeated viewings of 9 hours of videotape collected in one day-care center, they developed some hypotheses about hitting and a coding scheme they thought useful for children of those ages. The next step was to have a single observer collect data "live" in another day-care center. There were two reasons for this. First, given a well-worked-out coding scheme, they thought observing live would be more efficient (no videotapes to code later), and second, nursery school personnel were concerned about the disruption to their program that multiple observers and/or video equipment might entail. Two or more observers, each observing at different times, could have been used, but Brownlee and Bakeman thought that using one observer for the entire study would result in more consistent data. Further, the amount of observation required could easily be handled by one person. Nonetheless, two observers were trained, and agreement between them was checked before the "main" observer began collecting data. This was done so that the investigators, and others, would be convinced that this observer did not have a unique personal vision and that, on a few occasions at least, he and another person independently reported seeing essentially the same events.

Calibration

Just as assuring accuracy is the major conceptual reason, so calibrating observers is probably the major practical reason for establishing interobserver agreement. A study may involve a large number of separate observations and/or extend over several months or years. Whatever the reason, when different observers are used to collect the same kind of data, we need to assure ourselves that the data collected do not vary as a function of the observer. This means that we need to calibrate observers with each other or, better yet, calibrate all observers against some standard protocol.

Reliability decay

Not only do we need to assure ourselves that different observers are coding similar events in similar ways, we also need to be sure that an individual observer's coding is consistent over time. Taplin and Reid (1973) conducted a study of interobserver reliability as a function of observer's awareness that their coding was being checked by an independent observer. There were three groups – a group that was told that their work would not be checked, a group that was told that their work would be spot-checked at regular intervals, and a group that was told that their work would be randomly checked. Actually the work of all three groups was checked for all seven sessions. All groups showed a gradual decay in reliability from the 80% training level. The no-check group showed the largest decay. The spot-check group's reliability increased during sessions 3 and 6, when they thought they were being checked. The random-check group performed the best over all sessions, though lower than the spot-check group on session 3 and 6.

Reliability decay can be a serious problem when the coding process takes a long time, which is often the case in a large study that employs a complex coding scheme. One solution to the problem was reported by Gottman (1979a). Gottman obtained a significant increment in reliability over time by employing the following procedure in coding videotapes of marital interaction. One employee was designated the "reliability checker"; the reliability checker coded a random sample of *every* coder's work. A folder was kept for each coder to assess consistent confusion in coding, so that retraining could be conducted during the coder's periodic meetings with the reliability checker. To test for the possibility that the checker changed coding style for each coder, two procedures were employed. First, in one study the checker did not know who had been assigned to any particular tape until after it was coded. This procedure did not alter reliabilities. Second, coders occasionally served as reliability checkers for one another in another study. This procedure also did not alter reliabilities. Gottman also conducted a few studies that varied the amount of interaction that the checker coded. The reliabilities were essentially unaffected by sampling larger segments, with one exception: The reliabilities of infrequent codes are greatly affected by sampling smaller segments. It is thus necessary for each coding system to determine the amount that the checker codes as a function of the frequency of the least frequent codes.

What should be clear from the above is that investigators need to be concerned not just with inter-, but also with intraobserver reliability. An investigator who has dealt with the problems of inter- and intraobserver agreement especially well is Gerald Patterson, currently of the Oregon Social Learning Center. Over the past several years, Patterson and his co-workers have trained a number of observers to use their coding schemes.

Although observers record data live, training and agreement assessments depend on the use of videotapes. First, presumably "correct" codings or "standard" versions were prepared for a number of videotaped sequences. Then these standards were used to train new observers, who were not regarded as trained until their coding reached a preset criterion of accuracy, relative to the standard. Second, observers periodically recoded standard versions, and their agreement both with the standard and with their own previous coding was assessed. Such procedures require time and planning, but there is probably no other way to ensure the continued accuracy of human observers when a project requires more than one or two observers and lasts for more than a month or two.

4.2 Reliability versus agreement

So far in this chapter we have used the term "observer agreement," yet the term "observer reliability" often occurs in the literature. What, if any, is the difference? Johnson and Bolstad (1973) make a nice distinction. They argue that agreement is the more general term. It describes, as the word implies, the extent to which two observers agree with each other. Reliability is the more restrictive term. As used in psychometrics, it gauges how accurate a measure is, how close it comes to "truth." Hence when two observers are just compared with each other, only agreement can be reported. However, when an observer is compared against a standard protocol assumed to be "true," then observer reliability can be discussed.

Others would argue that reliability is the more general term. Interobserver agreement only addresses potential errors among observers and ignores many other sources of potential errors, which in the context of observational research may be many (Pedhazur & Schmelkin, 1991, pp. 114–115, 145–146). Yet, when two observers independently agree, the usual presumption is that they are therefore accurate, even though it is possible, of course, that they simply share a similar but nonetheless deviant worldview.

But this presumption is questionable because other sources of error may be present. In observational research (and in this book as well), interobserver agreement is emphasized, but it is important to remember that, although important, indices of interobserver agreement are not indices of reliability, and that reliability could be low even when interobserver agreement is high.

Not wishing to shortchange a complex topic (i.e., assessment of reliability), we would nonetheless argue that there is some merit in preparing standard protocols, presumed true, which can then be used as one, simple index of *observer reliability*. If the first reason for assessing observer agreement is to assure others that our observers are accurate and our procedures replicable, and the second reason is to calibrate multiple observers, then a third reason is to assure ourselves as investigators that observers are coding what we want (i.e., are seeing the world as we do). And one way to test this is to let observers code independently a sequence of events for which we have already prepared a standard protocol. Clearly, this is easiest to do when videotaped behavior or transcripts of conversations are coded, and less easy to do when live behavior is coded. However, such procedures let us speak, albeit relatively informally, of "observer reliability" (assuming of course that investigators are relatively infallible), and they also give investigators additional confidence in their observers, a confidence that probably becomes more important, the more socially based the coding scheme is.

In sum, "reliability" invokes a rich and complex psychometric tradition and poses problems that lie well beyond the scope of this book. From this point of view, "interobserver agreement" is the more limited and straightforward term. Moreover, it is the one that has been emphasized in observational research and, consequently, is addressed in the remainder of this chapter. The question now is, how should observer agreement be assessed and computed?

4.3 The problem with agreement percentages

Perhaps the most frequently encountered, and at the same time the most misleading, index of observer agreement is a percentage of some sort. This is usually referred to as a "percentage of agreement" and in its most general form is defined as follows:

$$P_A = \frac{N_A}{N_A + N_D} \times 100$$

P_A refers to the percentage of agreement, N_A the number of agreements, and N_D the number of disagreements. In any given application, the investigator would need to specify further the recording unit used (events or intervals), which is after all the basis for determining agreement and disagreement, and exactly how agreement and disagreement are defined.

For example, Tuculescu and Griswold (1983), in their study of prehatched chickens (section 2.10), defined four kinds of embryonic distress calls (Phioo, Soft Peep, Peep, and Screech). Observers coded events. Whenever an event of interest occurred, they recorded which it was, when it began, and when it ended. Given this coding scheme and this recording strategy, observer agreement could have been computed as follows.

First, what constitutes an agreement needs to be defined. For example, we might say that two observers agree if they record the same kind of distress call at times that either overlap or are separated by no more than two seconds. They disagree when one records a distress call and the other does not, or when they agree that a distress call occurred but disagree as to what kind it is. (Following an old classification system for sins, some writers call these disagreements "omission errors" and "commission errors," respectively.)

Once agreements and disagreements have been identified and tallied, the percentage of agreement can be computed. For example, if two observers both recorded eight Phioos at essentially the same time, but disagreed three times (each observer recorded one Phioo that the other did not, and once one observer recorded a Phioo that the other called a Soft Peep), the percentage of agreement would be 73 (8 divided by 8 + 3 times 100). Percentage agreement could also be reported, not just for Phioos in particular, but for embryonic distress calls in general. For example, if the two observers agreed as to type of distress call 35 times but disagreed 8 times, then the percentage of agreement would be 81 (35 divided by 35 + 8 times 100).

Given a reasonable definition for agreement and for disagreement, the percentage of agreement is easy enough to compute. However, it is not at all clear what the number means. It is commonly thought that agreement percentages are "good" if they are in the 90s, but there is no rational basis for this belief. The problem is that too many factors can affect the percentage of agreement – including the number of codes in the code catalog – so that comparability across studies is lost. One person's 91% can be someone else's 78%.

Perhaps the most telling argument against agreement percentage scores is this: Given a particular coding scheme and a particular recording strategy, some agreement would occur just by chance alone, even with blindfolded observers, and agreement percentage scores do not correct for this. This becomes most clear when an interval coding strategy is coupled with a simple mutually exclusive and exhaustive scheme, as in the study of parallel play described in section 1.7. Recall that for this study, Bakeman and Brownlee had observers code each successive 15-second interval as either Unoccupied, Solitary, Together, Parallel, or Group. If two observers had each coded the same 100 intervals, the pattern of agreement might have been as depicted in Figure 4.1. In this case, the percentage of agreement would be 87 (87 divided by 87 + 13 times 100). However, as we show in the next section, an agreement of 22.5% would be expected, in this case, just by chance alone. The problem with agreement percentages is that they do not take into account the part of the observed agreement that is due just to chance.

Figure 4.1 is sometimes called a "confusion matrix," and it is useful for monitoring areas of disagreement that are systematic or unsystematic. After computing the frequencies of entries in the confusion matrix, the reliability checker should scan for clusters off the diagonal. These indicate

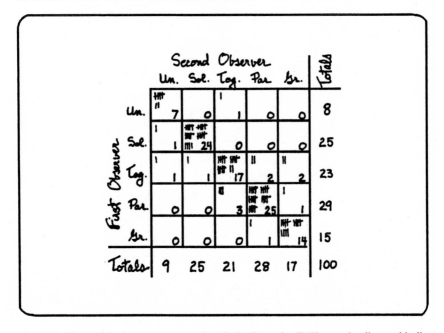

Figure 4.1. An agreement or "confusion" matrix. Tallies on the diagonal indicate agreement between the two observers, whereas tallies off the diagonal pinpoint disagreements.

confusions between specific codes. If many observers display the same confusion, it may suggest retraining, or clarification of the coding manual, or finding clear examples that sharpen distinctions between codes.

4.4 The advantages of Cohen's kappa

An agreement statistic that does correct for chance is Cohen's kappa (Cohen, 1960). As a result, it is almost always preferable to simple agreement percentages. It is defined as follows:

$$\kappa = \frac{P_{obs} - P_{exp}}{1 - P_{exp}}$$

where P_{obs} is the proportion of agreement actually observed and P_{exp} is the proportion expected by chance. P_{obs} is computed by summing up the tallies representing agreement (those on the upper-left, lower-right diagonal in Figure 4.1) and dividing by the total number of tallies. That is, it is analogous to P_A in the previous section, except that it is not multiplied by

100. Symbolically:

$$P_{obs} = \frac{\sum_{i=1}^{k} x_{ii}}{N}$$

where k is the number of codes (i.e., the order of the agreement matrix), x_{ii} is the number of tallies for the ith row and column (i.e., the diagonal cells), and N is the total number of tallies for the matrix. For the agreement portrayed in Figure 4.1, this is:

$$P_{obs} = \frac{7 + 24 + 17 + 25 + 14}{100} = .87$$

P_{exp} is computed by summing up the chance agreement probabilities for each category. For example, given the data in Figure 4.1, the probability that an interval would be coded Unoccupied was .08 for the first observer and .09 for the second. From basic probability theory, the probability of two events occurring jointly (in this case, both observers coding an interval Unoccupied), just due to chance, is the product of their simple probabilities. Thus the probability that both observers would code an interval Unoccupied just by chance is .0072 (.08 × .09). Similarly, the chance probability that both would code an interval Solitary is .0625 (.25 × .25), Together is .0483 (.21 × .23), Parallel is .0812 (.28 × .29), and Group is .0255 (.17 × .15). Summing the chance probabilities for each category gives the overall proportion of agreement expected by chance (P_{exp}), which in this case is .2247.

A bit of algebraic manipulation suggests a somewhat simpler way to compute P_{exp}. Multiply the first column by the first row total, add this to the second column total multiplied by the second row total, etc., and then divide the resulting sum of the column-row products by the total number of tallies squared. Symbolically:

$$P_{exp} = \frac{\sum_{i=1}^{k} x_{+i} x_{i+}}{N^2}$$

where x_{+i} and x_{i+} are the sums for the ith column and row, respectively (thus one row by column sum cross-product is computed for each diagonal cell).

For the agreement given in Figure 4.1, this is:

$$P_{exp} = \frac{9 \times 8 + 25 \times 25 + 21 \times 23 + 28 \times 29 + 17 \times 15}{100 \times 100}$$
$$= .2247$$

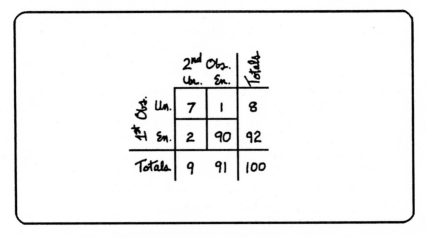

Figure 4.2. An agreement matrix using a coding scheme with two codes. Although there is 97% agreement, 84% would be expected just by chance alone.

Now we can compute kappa for our example data.

$$\kappa = \frac{.87 - .2247}{1. - .2247} = .8323 \text{ (rounded)}$$

As the reader can see, the amount of agreement corrected for chance (about .83) is rather less than the uncorrected value (.87). In some cases, especially when there are few coding categories and when the frequency with which those codes occur is quite disproportionate, the difference can be quite dramatic. Imagine, for example, that instead of the five categories listed in Figure 4.1, only two had been used: Unengaged (meaning Unoccupied) and Engaged. In that case, the data from Figure 4.1 could reduce to the data shown in Figure 4.2. The proportion of agreement oberved here is quite high, .97 (7 + 90 divided by 100), but so is the proportion of chance agreement as well.

$$P_{exp} = \frac{9 \times 8 + 91 \times 92}{100 \times 100} = .8444$$

As a result, the value of kappa, although still respectable, is considerably lower than the level of agreement implied (misleadingly) by the .97 value:

$$\kappa = \frac{.97 - .8444}{1. - .8444} = .8072$$

The question now is, is a kappa of .8072 big enough? Fleiss, Cohen, and Everitt (1969) have described the sampling distribution of kappa, and so it is possible to determine if any given value of kappa differs significantly from zero (see also Hubert, 1977). The way it works is as follows: First, the population variance for kappa, assuming that kappa is zero, is estimated from the sample data. Then the value of kappa estimated from the sample data is divided by the square root of the estimated variance and the result compared to the normal distribution. If the result were 2.58 or bigger, for example, we would claim that kappa differed significantly from zero at the .01 level or better.

In this paragraph, we show how to compute the estimated variance for kappa, first defining the procedure generally and then illustrating it, using the data from Figure 4.2. The formula incorporates the number of tallies (N, in this case 100), the probability of chance agreement (P_{exp} in this case .8444), and the row and column marginals: p_{i+} is the probability that a tally will fall in the ith *row*, whereas p_{+j} is the probability that a tally will fall in the jth *column*. In the present case, $p_{1+} = .08$, $p_{2+} = .92$, $p_{+1} = .09$, and $p_{+2} = .91$. To estimate the variance of kappa, first compute

$$\sum_{i=1}^{k} = p_{i+} \times p_{+i} \times [1 - (p_{+i} + p_{i+})]^2$$

In the present case, this is:

$$.08 \times .09 \times [1 - (.09 + .08)]^2 + .92 \times .91 \times [1 - (.91 + .92)]^2$$
$$= .00496 + .57675 = .5817$$

Then add to it this sum:

$$\sum_{\substack{i=1 \\ i \neq j}}^{k} \sum_{j=1}^{k} = p_{i+} \times p_{+j} \times (p_{+i} + p_{j+})^2$$

In the present case, this is:

$$.08 \times .91 \times (.09 + .92)^2 + .92 \times .09 \times (.91 + .08)^2$$
$$= .07426 + .08115 = .1554$$

Next subtract P_{exp}^2. In the present case this is:

$$.8444^2 = .7130$$

and the result is:

$$.5817 + .1554 - .7130 = .0241$$

Finally, divide this result by $N \times (1 - P_{exp})^2$. This divisor is:

$$100 \times (1 - .8444)^2 = 2.421$$

and the final quotient is:

$$.0241 / 2.421 = .009955$$

This is the estimated variance for kappa, given the data in Figure 4.2. The z score is 8.091, the estimated kappa (.8072) divided by the square root of the variance (.09977). We would conclude that the agreement demonstrated in Figure 4.2 is significantly better than chance.

For many investigators, this will not be stringent enough. Just as correlation coefficients that account for little variance in absolute terms are often significant, so too, quite low values of kappa often turn out to be significant. This means only that the pattern of agreement observed is greater than would be expected if the observers were guessing and not looking. This can be unsatisfactory, however. We want from our observers not just better than chance agreement; we want *good* agreement. Our own inclination, based on using kappa with a number of different coding schemes, is to regard kappas less than .7, even when significant, with some concern, but this is only an informal rule of thumb. Fleiss (1981), for example, characterizes kappas of .40 to .60 as fair, .60 to .75 as good, and over .75 as excellent.

The computation of kappa can be refined in at least three ways. The first is fairly technical. Multiple observers may be used – not just two – and investigators may want a generalized method for computing kappa across the different pairs. In such cases, readers should consult Uebersax (1982); a BASIC program that computes Uebersax's generalized kappa coefficient has been written by Oud and Sattler (1984).

The second refinement is often useful, especially when codes are roughly ordinal. Investigators may regard some disagreements (confusing Unoccupied with Group Play, for example) as more serious than others (confusing Together with Parallel Play, for example). Cohen (1968) has specified a way of weighting different disagreements differently. Three $k \times k$ matrices are involved: one for observed frequencies, one for expected frequencies, and one for weights. Let x_{ij}, m_{ij}, and w_{ij} represent elements from these three matrices, respectively: then $m_{ij} = (x_{+j} \times x_{i+}) \div N$, and the w_{ij} indicate how seriously we choose to regard various disagreements. Usually the diagonal elements of the weight matrix are 0, indicating agreement (i.e., $w_{ii} = 0$ for $i = 1$ through k); cells just off the diagonal are 1, indicating some disagreement; cells farther off the diagonal are 2, indicating more serious disagreement; etc. For the present example, we might enter 4 in cells x_{15} and x_{51}, indicating that confusions between *unoccupied* and *group* are given more weight than other disagreements. Then weighted kappa is computed as follows:

$$\kappa_{wt} = 1 - \frac{\displaystyle\sum_{i=1}^{k}\sum_{j=1}^{k} = w_{ij}x_{ij}}{\displaystyle\sum_{i=1}^{k}\sum_{j=1}^{k} w_{ij}m_{ij}}$$

If $w_{ii} = 0$ for $i = 1$ through k and 1 otherwise, then κ, as defined earlier, and κ_{wt} are identical. (To compute variance for weighted kappa, see Fleiss, Cohen, & Everitt, 1969).

A third way the computation of kappa can be refined involves P_{exp}. As usually presented, P_{exp} is computed from the row and column totals (the marginals), but in some cases investigators may believe they have better ways to estimate the expected distribution of the codes, and could use them to generate expected frequencies.

For example, imagine that an investigator has accumulated a considerable data archive using the coding scheme presented in Figure 4.1 and so has a good idea of how often intervals are coded Unoccupied, etc., in general. Imagine further that coders are subject to random agreement checks and that, as luck would have it, a tape selected for one such check shows an unusual child who spends almost all of his time Unoccupied with just a little additional Solitary play. In this case, the kappa would be quite low. What is really a 5-category scheme becomes for kappa computation a 2-category scheme with a skewed distribution, which usually results in low values. In effect, the observers have received no "credit" for knowing that Together, Parallel, and Group Play did not occur. In such a case, it may make sense to substitute the average values usually experienced for the marginals rather than use the actual ones from such a deviant case.

As the above example makes clear yet again, assessing observer agreement has more than one function. When our concern is to convince others (especially journal editors and reviewers) that our observers are accurate, then we might well pool tallies from several different agreement checks into one kappa table, computing and reporting a single kappa value. This has the merit of providing realistic marginals. When our concern is to calibrate and train observers, however, we would want to compute kappas separately for each agreement check, thus providing immediate feedback. At the same time, we should counsel our observers not to be discouraged when low kappas result simply from coding an unusual instance.

With respect to the training of observers, there is one final advantage of Cohen's kappa that we should mention. The kappa table itself provides a graphic display of disagreement. Casual inspection immediately reveals which codes are often confused and which almost never are. An excessive number of tallies in an off-diagonal cell would let us know, for example, that intervals one observer codes together are regarded as parallel by the other observer. Moreover, simple inspection can also reveal if one observer is more "sensitive" than the other. If so, a pattern of considerably more tallies above than below the diagonal (or vice versa) results. For example, if most of the disagreements in Figure 4.1 had been above the diagonal (meaning that what the first observer regarded as Unoccupied the second observer often regarded as something more engaged but not vice versa), this would indicate that the second observer was more sensitive, detecting engagement when the first observer saw only an unengaged child. Such patterns have clear implications for observer training. In the first case (code confusion), further training would attempt to establish a consensual definition for the two codes; and in the second (different sensitivities), further training would attempt to establish consensual thresholds for all codes.

4.5 Agreement about unitizing

Constructing a kappa table like that shown in Figure 4.1 is easy enough. For each "unit" coded, a tally is entered in the table. The codes assigned the unit by the two observers determine in which cell the tally is placed. This procedure, however, assumes previous agreement as to what constitutes a unit. Sometimes unit boundaries are clear-cut, and may even be determined by forces external to the observers. At other times, boundaries are not all that distinct. In fact, determining unit boundaries may be part of the work observers are asked to do. For example, observers may be asked to identify and code relatively homogeneous stretches of talk, or to identify a particular kind of episode (e.g., a negotiation or conflict episode).

In such cases, agreement needs to be demonstrated on two levels: first with respect to "unitizing" (i.e., identifying the homogeneous stretches or episodes), and second with respect to the codes assigned the homogeneous stretch or episode (or perhaps events embedded within the episode itself).

When the unit being coded is a time interval, as in the example just given, there is no problem. In that case, unit boundaries are determined by a clock and not by observers. When the unit being coded is an event, however, the matter becomes more difficult. For example, consider Gottman's study of friendship formation described in section 2.11. In such cases, there are two parts to the coding task. First, observers need to segment the stream of recorded talk into thought units, whereas second, they need to code the segmented thought units themselves.

Agreement with respect to the coding of thought units can be determined using Cohen's kappa, as described in the last section. However, how should agreement with respect to segmenting the stream of talk into thought units be demonstrated? In the older literature, a percentage score has often been used for this task.

If both observers worked from transcripts, marking thought unit boundaries on them, it should be a fairly simple matter to tally the boundaries claimed by both observers and the ones noted only by one observer. But in this case, there can be only omission disagreements. Moreover the percentage agreement would not correct for chance; still the score may have some descriptive value, albeit limited.

Alternatively, the investigator might regard every gap between adjacent words as a potential boundary. In this case, the initial coding unit would be the word gap. Each gap would contribute one tally to a 2 × 2 kappa table: (a) Both observers agree that this gap is a thought unit boundary; (b) the first observer thinks it is, but the second does not; (c) the second thinks it is, but the first disagrees; or (d) both agree that it is not a thought unit boundary. Although the chance agreement would likely be high (two coding categories, skewed distribution, assuming that most gaps would not be boundaries), still the kappa computed would correct for that level of chance agreement.

A better and more general procedure for determining agreement with respect to unitizing (i.e., identifying homogeneous stretches of talk or particular kinds of episodes) requires that onset and offset times for the episodes be available. Then the tallying unit for the kappa table becomes the unit used for recording time. For example, imagine that times are recorded to the nearest second and that observers are asked to identify conflict episodes, recording their onset and offset times. Then kappa is computed on the basis of a simple 2 × 2 table like that shown in Figure 4.2, except that now rows and columns are labeled yes/no, indicating whether or not a second was coded for conflict. One further refinement is possible. When tallying seconds, we could place a tally in the agreement (i.e., yes/yes) cell if one observer claimed conflict for the second and the other observer claimed conflict either for that second or an adjacent one, thereby counting 1-second disagreement as agreements as often seems reasonable.

In the previous chapter, we described five general strategies for recording observational data: (a) coding events, (b) timing onsets and offsets, (c) timing pattern changes, (d) coding intervals, and (e) cross-classifying events. Now we would like to mention each in turn, describing the particular problems each strategy presents for determining agreement about unitizing.

Coding events, without any time information, is in some ways the most problematic. If observers work from transcripts, marking event (thought unit) boundaries, then the procedures outlined in the preceding paragraphs can be applied. If observers note only the sequence of events, which means that the recorded data consist of a string of numbers or symbols, each representing a particular event or behavioral state, then determining agreement as to unit boundaries is more difficult. The two protocols would need to be aligned, which is relatively easy when agreement is high, and much more difficult when it is not, and which requires some judgment in any case. An example is presented in Figure 4.3.

When onset and offset or pattern-change times are recorded, however, the matter is easier. Imagine, for example, that times are recorded to the nearest second. Then the second can be the unit used for computing agreement both for unitizing (identifying homogeneous stretches or episodes) and for the individual codes themselves. Because second boundaries are determined

1st Obs.	2nd Obs.	Method A	Method B
U	U	a	a
S	S	a	a
G	G	a	a
T	P	d	d
G	G	a	a
P			d
S	S	a	a
P	T	d	d
U	U	a	a
T			d
U			a
P	P	a	a
G	G	a	a

Figure 4.3. Two methods for determining agreements ("a") and disagreements ("d") when two observers have independently coded the same sequence of events. Method A ignores errors of omission. Method B counts both errors of commission and errors of omission as disagreements.

by a clock external to the observers, there is no disagreement as to where these boundaries fall (the only practical requirement is that the clocks used by the two observers during an agreement check be synchronized in some way). An example showing how second-by-second agreement would be computed in such cases is given in the next section.

When time intervals are coded in the first place, the matter is similar. Again, the underlying unitization is done by clocks, not by observers. When cross-classifying events, however, we need to ask, to what extent are both observers detecting the same events? In older literature, often a percentage agreement was used. For example, in their study of social rules among preschool children, Bakeman and Brownlee (1982) asked two observers to cross-classify object struggles (see section 2.13). During an agreement check, one observer recorded 50 such struggles, the other 44; however, all 44 of the latter had also been noted by the first observer, and hence their percentage agreement was 88.0% (44 divided by 44 + 6). In a case like this, there seems to be no obvious way to correct for chance agreement.

Our recommendation is as follows: Report the agreement-disagreement tallies along with the percentage of agreement in such cases, but note their limitations. However, if at all possible, report time-based kappa statistics to establish that observers detected the same events to cross-classify. This is the same strategy we recommend to demonstrate that observers are identifying the same homogeneous stretches or episodes, which likewise are then subjected to further coding.

4.6 Agreement about codes: Examples using Cohen's kappa

In this section, we describe how Cohen's kappa can be used to document observer agreement about codes for each of the recording strategies listed in chapter 3. This is not the only way to determine observer agreement (or reliability), as we discuss in the next section, but it may be among the most stringent. This is because Cohen's kappa documents point-by-point agreement, whereas many writers would argue that agreement does not need to be determined for a level more detailed than that ultimately used for analysis. We think that this argument has merit, but that there are at least two reasons to favor a relatively stringent statistic like Cohen's kappa. First, as we argued earlier in this chapter, determining observer agreement has more than one function. For training observers and providing them feedback on their performance, we favor an approach that demands point-by-point agreement. We also like the graphic information about disagreement provided by the kappa table. Second, once agreement at a detailed level has been established, we can safely assume agreement at less detailed levels, and in any case a relatively detailed level is required for sequential analysis.

When observers code events, it is relatively straightforward to compute agreement about how units are coded, once the units are identified. For example, when thought units are coded from transcripts, the codes the two observers assign each successive thought unit would determine the cell of a 26×26 kappa table in which a tally would be placed (assuming 26 possible codes for thought units). What this example highlights is that kappa is a summary statistic, describing agreement with respect to how a coding scheme is used (not agreement about particular codes in the scheme), and that the codes that define each kappa table must be mutually exclusive and exhaustive.

As a second example, still assuming that events are being coded, imagine that an investigator wants to segment the stream of behavior into the five play states used by Bakeman and Brownlee (Unoccupied, Solitary, Together, Parallel, Group). If observers are told where the segment boundaries are, their only task would be to code segments. In this case, it would be easy to construct a kappa table. Telling observers where the segment boundaries are, however, is not an easy matter. Other observers, working with videotaped material, would need to have determined those boundaries previously, and then the boundaries would need to be marked in some way, perhaps with a brief tone dubbed on the soundtrack.

If observers are not told where the segment boundaries are, however, but instead are asked to both segment and code at the same time, then protocols like those shown in Figure 4.3 would result. The question now is, how do we construct a kappa table from data like these? There are two choices. We could ignore those parts of the protocols where observers did not agree as to segment boundaries and tally only those segments whose boundaries were agreed upon, in effect ignoring errors of omission. This would result in eight agreements and two disagreements, as shown in Figure 4.3 (Method A). Or we could assume that there "really" was a segment boundary whenever one of the observers said there was. This would result in nine agreements and four disagreements (Method B). Neither of these choices seems completely satisfactory. The first probably overestimates agreement, whereas the second probably underestimates it. Our preference is for computing kappas using a time interval as the unit, but this requires timing onsets and offsets, timing pattern changes, or coding intervals directly.

As an example, consider the recording of onset and offset times for different kinds of embryonic distress calls done by Tuculescu and Griswold (section 2.10). Assume that these times were recorded to the nearest second. Then each second can be categorized as "containing" (a) a Phioo, (b) a Soft Peep, (c) a Peep, (d) a Screech, or (e) no distress call. If two observers both code the same audiotape, agreement data could be like that shown in Figure 4.4. Similarly, in addition to the kappa for embryonic distress calls, kappas could be computed for Tuculescu and Griswold's other superordinate categories as well (embryonic pleasure calls, maternal body movements, maternal head movements, and maternal vocalizations).

Occasionally, editors or colleagues ask for agreement statistics, not for a coding scheme, as kappa provides, but for individual codes. We usually think that kappa coupled with the agreement matrix is sufficient, but nonetheless kappas can be computed for individual codes by collapsing the table appropriately. Consider the agreement matrix for the five distress calls shown in Figure 4.4. From it we could derive five 2×2 matrices, first collapsing the tallies into Phioo/not Phioo, then Soft Peep/not Soft Peep, etc. Then a kappa could be computed separately for each table. In this case, the individual kappas would be .79, .87, .43, .98, and .93 for Phioo, Soft Peep, Peep, Screech, and None, respectively. Not surprisingly, the kappa associated with Peep is relatively low; of the 10 noted by the first observer and the 13 noted by the second observer, agreement occurred for only 5.

A second example of a time-based kappa is provided by Adamson and Bakeman (1985), who asked observers to record whenever infants displayed heightened affectivity. These displays were typically quite brief,

		Second Observer					
		Phioo	Soft Peep	Peep	Screech	None	Totals
First Observer	Phioo	42	3	0	0	8	53
	Soft Peep	1	93	7	0	5	106
	Peep	2	3	5	0	0	10
	Screech	0	1	0	162	2	165
	None	7	4	1	1	430	443
	Totals	52	104	13	163	445	777

Figure 4.4. An agreement matrix for coding of chicken embryonic distress calls. Each tally represents a 1-second interval. The reader may want to verify that the percentage of agreement observed is 94.2%, the percentage expected by chance is 39.4%, the value of kappa is .904, its standard error is .0231, and the z score comparing kappa to its standard error is 39.2.

lasting just a few seconds, relatively infrequent, and consisted of such things as smiles, gleeful vocalizations, or excited arm waving. Assuming a unit of 1 second, a 10-minute agreement check could produce a kappa table like the one given on the left in Figure 4.5. What would happen, however, if a half-second unit had been used instead? The answer is essentially nothing, as is demonstrated by the table on the right in Figure 4.5. By and large, halving the length of the unit would result in tables that are roughly proportional, except that one would have twice as many tallies as the other. The kappa statistic (unlike chi-square) is not affected by this, however, as the kappa computations in Figure 4.5 demonstrate.

In the preceding two paragraphs (and in Figures 4.4 and 4.5), we have suggested how kappa can be computed when onset times for mutually exclusive and exhaustive codes are recorded. The principle is exactly the same when timing pattern changes. Again, for each set of mutually exclusive and exhaustive codes, a kappa table can be constructed, tallying agreement and disagreement for each second (or whatever time unit is used) coded. When timing pattern changes, codes are always parceled into

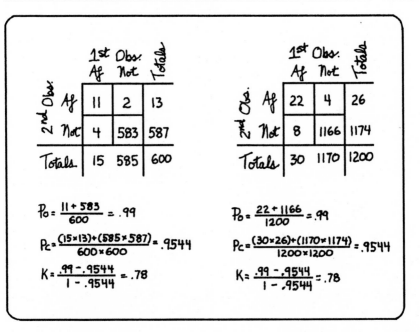

Figure 4.5. When time intervals are tallied, and a reasonable time interval is used, the value of kappa is not changed by halving the time interval (which doubles the tallies).

mutually exclusive and exhaustive sets. In other cases, constructing a set of mutually exclusive and exhaustive codes is no problem, as the examples presented in Figures 4.4 and 4.5 demonstrate. Adding the "none of the above" code to a code for affective display or to codes for distress calls makes the set mutually exclusive and exhaustive.

When the remaining two recording strategies – coding intervals and cross-classifying events – are used, computing agreement using kappa is straightforward. An example of what a kappa table might look like when intervals are coded was given earlier (see Figure 4.1). When events are cross-classified, the only difference is that there is one kappa table for each classification scheme (or dimension) and that events not detected by both observers cannot be entered into the table. For example, in their study of social rules, Bakeman and Brownlee (1982) reported kappas for each of their three dimensions: (a) prior possession, (b) resistance, and (c) success (see section 2.13). Kappa, however, is not the only agreement statistic there is. As we discuss in the next section, other statistics and approaches to observer agreement have their advantages.

4.7 Generalizability theory

Cronbach, Gleser, Nanda, and Rajaratnam (1972; see also Brennan, 1983) presented what amounts to a conceptual breakthrough in thinking about both reliability and validity. To understand their notions, let us introduce the concept of the "work we want the measure to do." For example, we would like to be able to use our observations of the amount of negative-affect reciprocity to discriminate satisfied from dissatisfied marriages. Or, we might want our measure of the amount of negative affect to predict the husband's health in three years. This is the work our measure is to do. It is designed to discriminate or to predict something of interest.

Cronbach and colleagues' major point is that this work is always relative to our desire in measurement to generalize across some facet of our experimental design that we consider irrelevant to this work. For example, our test scores should generalize across items within a measurement domain. It should not matter much if we correlate math achievement and grade point average (GPA) using even or odd math achievement items to compute the correlation coefficient. We are generalizing across the irrelevant facet of odd/even items. The work the measure does is to discriminate high- from low-GPA students.

In a similar way, if we have a measure of negative affect, we expect it to discriminate among happily and unhappily married couples, and not to discriminate among coders (the irrelevant facet). We wish to generalize across coders. Let us briefly discuss the computations involved in this analysis. Figure 4.6 presents the results of one possible generalizability study. For five persons, each of two observers computed the frequency of code A for a randomly selected segment of videotape. The setup is the same as a simple repeated-measures experiment. The within-subject factor is observer (with two levels, i.e., data from two observers) and there is no between-subject factor as such; subjects represent total between-subject variability. The analysis of variance source table for the data shown in Figure 4.6, expanded to include R^2 as recommended by Bakeman (1992), is shown in Table 4.1. Given these data and the current question, an appropriate coefficient of generalizability, or reliability, is:

$$\alpha = \frac{MS_p - MS_r}{MS_p + (n_o - 1)MS_r} \tag{4.1}$$

where n_o is the number of observers (2 in this case) and MS_p and MS_r are the mean squares for persons and residual (or error), respectively (the first edition of this book omitted $n_o - 1$ in the denominator because it equaled 1, but this proved confusing). This is an intraclass correlation coefficient based on the classical assumption that observed scores can be divided into a true and an error component ($X = T + e$), so that the appropriate intraclass

Code A's Frequency

Person	Observer 1	Observer 2	Person Average
1	2	1	$\bar{P_1} = 1.5$
2	20	14	$\bar{P_2} = 17.0$
3	30	22	$\bar{P_3} = 26.0$
4	3	7	$\bar{P_4} = 5.0$
5	120	84	$\bar{P_5} = 102.0$
	$\bar{O_1} = 35.0$	$\bar{O_2} = 25.6$	$\bar{M} = 30.3$ (grand mean)

$$MS_p = \frac{1}{5-1} n_o \sum_{p=1}^{n_p} \left(\bar{P_p} - \bar{M}\right)^2 = 3402.9$$

$$MS_o = \frac{1}{2-1} n_p \sum_{i=1}^{n_o} \left(\bar{O_i} - \bar{M}\right)^2 = 220.9$$

$$MS_r = \frac{1}{(5-1)(2-1)} \sum_{p=1}^{n_p} \sum_{i=1}^{n_o} \left(X_{pi} + \bar{M} - \bar{P_p} - \bar{O_i}\right)^2 = 121.4$$

$$\alpha = \frac{MS_p - MS_r}{MS_p + (n_o - 1) MS_r} = 0.93$$

Figure 4.6. A generalizability approach to observer agreement: n_o is the number of observers or 2, n_p is the number of persons or 5, MS_p is the mean square for persons, MS_o is the mean square for observers, and MS_r is the mean square for residual or, in this case, the $P \times O$ interaction; see also Table 4.1. The formulas for MS_p and MS_o were incorrect in the first edition of this book; they are also given incorrectly in Wiggins (1973, p. 289).

correlation is defined as

$$\alpha = \frac{\sigma_T^2}{\sigma_T^2 + \sigma_e^2}. \tag{4.2}$$

(Equation 4.1 is derived from 4.2 by substitution and algebraic manipulation.) This statistic estimates the reliability of observations made by a randomly selected observer, selected from the pool that contained the two observers used here for the reliability study (Table 4.1), and further

Table 4.1. *Source Table for the Generalizability Study of Observer Reliability*

Source	R^2	ΔR^2	SS	df	MS
Person	0.95	0.95	13,611.6	4	3,402.9
Observer	0.97	0.02	220.9	1	220.9
P × O	1.00	0.03	485.6	4	121.4
Total			14,318.1	9	

assumes that data will be interpreted within what Suen (1988) terms a norm-referenced (i.e., values are meaningful only relatively; rank-order statistics like correlation coefficients are emphasized) as opposed to a criterion-referenced framework (i.e., interpretation of values references an absolute external standard; statistics like unstandardized regression coefficients are emphasized). Equation 4.1 is based on recommendations made by Hartmann (1982) and Wiggins (1973, p. 290). For other possible intraclass correlation coefficients (generalizability coefficients), based on other assumptions, see Fleiss (1986, chapter 1) and Suen (1988), although, as a practical matter, values may not be greatly different. For example, values for a criterion-referenced fixed-effect and random-effect model per Fleiss (1986) were .926 and .921, respectively, compared to the .931 of Figure 4.6. In contrast, assuming that observers were item scores and we wished to know the reliability of total scores based on these items, Cronbach's internal-consistency alpha (which is $MS_p - MS_r$ divided by MS_p; see Wiggins, 1973, p. 291) was .964.

This way of thinking has profound consequences. It means that reliability can be high even if interobserver agreement is moderate, or even low. How can this be? Suppose that for person #5 in Figure 4.6, Observer 1 detected code A 120 times, as shown but only 30 of these overlapped in time with Observer 2's 84 entries. Then the interobserver agreement would be only $30/120 = .25$. Nonetheless, the generalizability coefficient of equation 4.1 is .93. The reliability is high because either observer's data distinguishes equally well between persons. The agreement within person need not be high. The measure does the work it was intended to do, and either observer's data will do this work. This is an entirely different notion of reliability than the one we have been discussing.

Note that the generalizability or reliability coefficient estimated by equation 4.1 is a specific measure of the relative variance accounted for by an interesting facet of the design (subjects) compared to an uninteresting one (coders). This is an explicit and specific proportion, but it does not tell us how large a number is acceptable, any more than does the proportion of variance accounted for in a dependent variable by an independent variable. The judgement must be made by the investigator. It is not automatic, just as a judgment of an adequate size for kappa is not an automatic procedure.

Note also that what makes the reliability high in the table in Figure 4.6 is having a wide range of people in the data, ranging widely with respect to code A. Jones, Reid, and Patterson (1975) presented the first application of Cronbach and colleagues' (1972) theory of measurement to observational data.

The reliability analysis just presented, although appropriate when scores are interval-scaled (e.g., number of events coded A by an observer), is inadequate for sequential analysis. The agreement required for sequential analysis cannot be collapsed over time, but must match point for point, as exemplified by the kappa tables presented in the previous section. Such matching is much more consistent with "classical" notions of reliability, i.e., before Cronbach et al. (1972).

Still, agreement point-for-point could be assessed in the same manner as in Figure 4.6. The two columns in Figure 4.6 would be replaced by sums from the confusion matrix. Specifically, the sums on the diagonal would replace Observer 1's scores, and the sums of diagonal plus off-diagonal cells (i.e., the row marginals) would replace Observer 2's scores. If the agreement point-for-point were perfect, all entries for code A in the confusion matrix would be on the diagonal and the two column entries would be the same. There would then be no variation across "observers," and alpha would be high. In this case, the "persons" of Figure 4.6 become codes, "Observer 1" becomes agreements and "Observer 2" becomes agreements plus disagreements.

This criterion is certainly sufficient for sequential analysis. However, it is quite stringent. Gottman (1980a) proposed the following: If independent observers produce similar indexes of sequential connection between codes in the generalizability sense, then reliability is established. For example, if two observers produced the data in Figure 4.7 (designed so that they are off by one time unit, but see the same sequences), their interobserver agreement would be low but indexes of sequential connection would be very similar across observers. Some investigators handle this simple problem by having a larger time window within which to calculate the entries in the confusion matrix. However, that is not a general solution because more complex configurations than that of Figure 4.7 are possible, in which both observers detect similar sequential structure in the codes but point-for-point agreement is low. Cronbach et al.'s (1972) theory implies that all we need to demonstrate is that observers are essentially interchangeable in doing the work that our measures need to do.

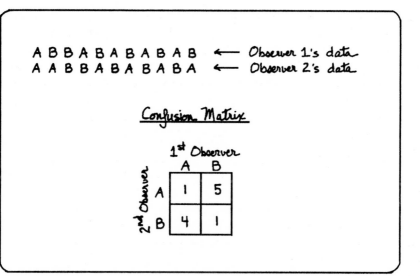

Figure 4.7. A confusion matrix when observers "see" the same sequence but one observer lags the other. In such cases, a "point-for-point" agreement approach may be too stringent.

4.8 Unreliability as a research variable

Raush (personal communication, 1974) once suggested that one source of unreliability is to be found in the nature of the social interaction itself. He referred to a message called a "probe" that a person might send to the receiver. The probe is designed to be taken one of two ways, depending on the state of the receiver. For example, in a potentially sexual interaction a sender may send a probe with a subtle sexual invitation. If it is ignored, the sender gains information that directs the interaction one way; if it is responded to, the interaction may proceed in another direction. Krokoff (1983) recently tested this notion in marital interaction. He reasoned that such probe messages would be more common during high-conflict interaction because of the delicate nature of the subject matter and the great danger that the conflict would escalate. If this were true, Krokoff reasoned, then reliability would be significantly reduced for those videotapes high in negative affect. This hypothesis was strongly supported by the data.

Patterson's (1982, p. 50) book quoted Reid as noting that "observer agreement is largely a function of the complexity of the interaction. By selecting *simple* interaction segments, one may obtain *very* high observer agreement." When complexity was defined as the number of different codes entered divided by the total entries for 5 minutes, this hypothesis was strongly supported; reliability was lower for more complex segments. This could partly be due to the increased task demands on the coder, but it could also be partly a property of the interaction itself, if in more complex interactions people sent more probe messages. The point of this section is to suggest that reliability can itself become a research variable of interest.

4.9 Summary

There are at least three major reasons for examining agreement among observers. The first is to assure ourselves and others that our observers are accurate and that our procedures are replicable. The second is to calibrate multiple observers with each other or with some assumed standard. This is important when the coding task is too large for one observer or when it requires more than a few weeks to complete. The third reason is to provide feedback when observers are being trained.

Depending on which reason is paramount, computation of agreement statistics may proceed in different ways. One general guiding principle is that agreement need be demonstrated only at the level of whatever scores are finally analyzed. Thus if conditional probabilities are analyzed, it is sufficient to show that data derived from two observers independently coding the same stream of behavior yielded similar conditional probabilities. Such an approach may not be adequate, however, when training of observers is the primary consideration. Then, point-by-point agreement may be demanded. Point-by-point agreement is also necessary when data derived from different observers making multiple coding passes through a videotape are to be merged later.

When an investigator has sequential concerns in mind, then, point-by-point agreement is necessary for observer training and is required for at least some uses of the data. Moreover, if point-by-point agreement is established, it can generally be assumed that scores derived from the raw sequential data (like conditional probabilities) will also agree. If agreement at a lower level is demonstrated, agreement at a higher level can be assumed. For these reasons, we have stressed Cohen's kappa in this chapter because it is a statistic that can be used to demonstrate point-by-point agreement. A Pascal program that computes kappa and weighted kappa is given in the Appendix. Kappa is also computed by Bakeman and Quera's Generalized Sequential Querier or GSEQ program (Bakeman & Quera, 1995a). At the same time, we have also mentioned other approaches to observer reliability (in section 4.7). This hardly exhausts what is a complex and far-ranging topic. Interested readers may want to consult, among others, Hollenbeck (1978) and Hartmann (1977, 1982).

5

Representing observational data

5.1 Representing versus recording

When observational data are being recorded, it is reasonable that practical concerns dictate what is done. One uses whatever equipment is available and convenient. One records in ways that are easy and natural for the human observers. One tries to preserve in some form the information that seems important. Thus recorded data can appear in literally a multitude of forms. Some of these were described in chapter 3; however, we recognize that what investigators actually do often combines the simple forms described there into more complex, hybrid forms.

However, the form used for data recording – the data as collected – should not become a straitjacket for the analysis to follow. A format that works well for data recording may not work so well for data analysis, and a format that works well for one analysis may not work well for another analysis. The solution is to figure out simple ways to represent data, ways that make different kinds of analysis simple and straightforward. There is nothing especially sacrosanct about the form of the recorded data, after all, and there is no merit in preserving that form when it proves awkward for subsequent uses.

It would be very useful if just a few relatively standard forms for representing observational data could be defined. Not only would this help to standardize terminology with respect to sequential analysis, thus facilitating communication, it would also make analysis easier and would facilitate designing and writing general-purpose computer programs for sequential analysis. We assume that most investigators use computers for their data analysis tasks, but even if they do not, we think that representing the data by use of one (or more) of the five standard forms defined in this chapter will make both thinking about and doing data analysis a simpler and more straightforward affair. Further, there is nothing exclusive about these five forms. Depending on how data were recorded, investigators can, and probably often will, extract different representations from the same recorded data for different purposes.

The first four forms are defined by Bakeman and Quera's (1992, 1995a) Sequential Data Interchange Standard (SDIS), which defines a standard form for event, state, timed-event, and interval sequences, respectively. Sequential data represented by any of these forms can be analyzed with the Generalized Sequential Querier (GSEQ; Bakeman & Quera, 1995a). The fifth form is an application of the standard cases by variables rectangular matrix and is useful for analyzing cross-classified events, including contingency table data produced by the GSEQ program.

5.2 Event sequences

The simplest way to represent sequential behavior is as event sequences. As an example, imagine that observers used the following mnemonic codes for the parallel play study described in chapter 1: Un = Unoccupied, Sol = Solitary, Tg = Together, Par = Parallel, and Gr = Group. A child is observed for just a minute or two. First she is unoccupied, then she shifts into together play, than back to unoccupied, back to together, then into solitary play, and back to together again. If each code represents a behavioral state, then the event sequence data for this brief observation would look like this:

Un   Tg   Un   Tg   Sol   Tg . . .

In this case, there are no Par or Gr codes because the child was not observed in parallel or group play during this observation session.

The data might have been recorded directly in this form or they might have been recorded in a more complex form and only later reduced to event sequences. The behavior to be coded could be thought of as behavioral states, as in this example, or as discrete events. Behavior other than just the states or events that form the event sequences might have been recorded but ignored when forming event sequences for subsequent analysis. In general, if the investigator can define a set of mutually exclusive and exhaustive codes of interest, and if the sequence in which those codes occurred can be extracted from the data as recorded, and if reasonable continuity between successive codes can be assumed (as discussed in section 3.4), then some (if not all) of the information present in the data can be represented as event sequences.

This is often very desirable to do. For one thing, the form is very simple. A single stream of codes is presented without any information concerning time, whether onsets or offsets. Lines of an event-sequential data file consist simply of codes for the various events, ordered as they occurred in time, along with information identifying the participant(s) and sessions. Thus event sequential data are appropriate when observed behavior is reduced to a single stream of coded events (which are thus mutually exclusive and exhaustive by definition), and when information about time (such as the duration of events) is not of interest. Event sequences are both simple and limited. Yet applying techniques described in chapter 7 to event sequences, Bakeman and Brownlee were able to conclude that parallel play often preceded group play among 3-year-old children.

5.3 State sequences

For some analyses, the duration of particular behavioral states or events may matter, either because the investigator wants to know how long a particular kind of event lasted on the average or because the investigator wants to know what proportion of the observation time was devoted to a particular kind of event. For example, it may be important to know that the mean length for a bout of parallel play was 34 seconds and that children spent 28% of their time engaged in parallel play, on the average. In such cases, the form in which data are represented needs to include information about how long each event or behavioral state lasted.

As we define matters, state sequences are identical to time sequences with the simple addition of timing information. The terminology is somewhat arbitrary, and in the next section we discuss timed-event sequences, but our intent is to provide a few simple forms for representing sequential data, some of which are simpler than others, so that investigators can choose a form that is no more complex than required for their work. For example, if duration were important, the SDIS state-sequential representation for the sequence given earlier would be:

$$Un = 12 \qquad Tg = 8 \qquad Un = 21 \qquad Tg = 11$$
$$Sol = 34 \qquad Tg = 6 \; \cdots$$

Assuming a time unit of a second, this indicates that Unoccupied lasted 12 seconds, followed by 8 seconds of Together, 21 more seconds of Unoccupied, etc. The same sequence can also be represented as:

$$Un,8{:}01 \qquad Tg,8{:}13 \qquad Un,8{:}21$$
$$Tg,8{:}42 \qquad Sol,8{:}53 \qquad Tg,9{:}27 \qquad ,9{:}33 \; \cdots$$

which indicates an onset time for Unoccupied of 8 minutes and 1 second, for the first Together of 8 minutes and 13 seconds, etc. The offset time for the session is 9 minutes and 33 seconds; because the first onset time was 8:01, the entire session lasted 92 seconds.

State sequences, like simple event sequences, provide a useful way to represent aspects of the recorded data when a single stream of mutually exclusive and exhaustive (ME&E) coded states (or events) captures information of interest. It would be used instead of (or in addition to) state sequences when information concerning proportion of time devoted to a behavior (e.g., percentage of time spent in parallel play) or other timing information (e.g., average bout length for group play) is desired. Additionally, it is possible to define multiple streams of ME&E states; for details see Bakeman and Quera (1995a). In sum, both simple event and state sequences are useful for identifying sequential patterns, given a simple ME&E coding scheme. But they are not useful for identifying concurrent patterns (unless multiple streams of states are defined) or for answering relatively specific questions, given more complex coding schemes. In such cases, the timed-event sequences described in the next section may be more useful.

5.4 Timed-event sequences

If codes can cooccur, and if their onset and offset times were recorded, then the data as collected can be represented as timed-event sequences. This is a useful and general-purpose format. Once data are represented in this form, an investigator can determine quite easily such things as how often specific behavioral codes cooccur (does the baby smile mainly when the mother is looking at him or her?), or whether certain behavioral codes tend to follow (or precede) other codes in systematic ways (does the mother respond to her baby's smiling within 5 seconds?).

For example, imagine that two people engaged in conversation were videotaped, and four codes were defined: *Alook,* meaning person A looks at person B; *Blook*, meaning B looks at A; *Atalk*, and *Btalk*. A brief segment of the coded conversation, depicted as though it had been recorded with an old-fashioned, deflected-pen, rolling-paper event recorder, is shown in Figure 5.1. Following SDIS conventions for timed-event sequences, this same segment would be represented as follows:

,1 Alook,2–4   Atalk,3–10   Blook,4–7   Alook,6
Blook,8–10   Alook,12–16   Btalk,12–19
Blook,13   Alook,17–20   Atalk18 , 21

Assuming time units are a second, this session began at second 1 and ended at second 21; thus the session lasted 20 seconds. Person A first began looking at second 2 and stopped at second 4; thus A's first looking bout lasted 2 seconds, etc. By convention, when offset times are omitted, durations are assumed to be one time unit; thus *Alook, 6* implies an offset time of 7. As you can see, offset times are assumed to be exclusive, but

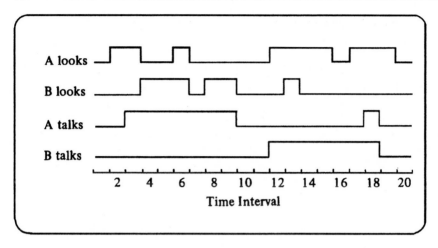

Figure 5.1. An example of the kind of output produced by a deflected-pen event recorder.

SDIS conventions also allow inclusive offset times (signaled by a right parenthesis). Thus the segment just given could also be represented as:

,1 Alook, 2–3)   Atalk, 3–9)   Blook, 4 – 6)   Alook, 6
Blook, 8–9)   Alook, 12–15)   Btalk, 12–18)
Blook, 13   Alook, 17–19)   Atalk, 18 ,20)

which some investigators may find more convenient. Other conventions for timed-event sequences, including inferred offset times and context codes, are detailed in Bakeman and Quera (1995a).

The flexibility of the timed-event format is very useful. As in other SDIS formats, codes can be defined for any number of behaviors (although, as a practical matter, the SDIS program limits the number to 95), but unlike event sequence data, timed-event sequential data (TSD) preserves onset and offset times, and unlike a single stream of state sequences, TSD preserves cooccurrences among behaviors. This allows for the level of complexity represented in the data to approach "real life," and it certainly imposes a minimum of constraints on the questions investigators can ask.

5.5 Interval sequences

The fourth format defined by SDIS, interval sequences, is designed to accommodate interval recording in a simple and straightforward way. Codes are simply listed as they occur and interval boundaries are represented by commas. For example:

Interval = 5;
, , Ivoc, Ivoc, Aofr, Rain Aofr Ivoc Ismi, Rain, . . .

indicates that intervals have a duration of 5 time units. No behavior is coded for the first two intervals. The infant vocalizes in intervals 3 and 4 and an adult offers in interval 5. The adult continues to (or again) offers in interval 6, at which time the infant both vocalizes and smiles and it begins to rain. The rain continues in interval 7, etc. Other conventions for interval sequences, including context codes, are detailed in Bakeman and Quera (1995a).

If the interval width for interval sequences were 1 second, for example, and if the time unit for timed-event sequences were likewise 1 second, then the same observed sequence could be represented with either timed-event sequences or interval sequences. Both allow behavioral codes to cooccur. But when an investigator begins with an interval recording strategy, it is often easier (and requires fewer key strokes) to represent such data directly as interval sequences, which is why we included this format as part of SDIS.

Many computer-based data collection devices and programs in common use produce data in their own format. For example, timed-event sequences in SDIS place the code first, followed by a comma, followed by the time the code occurred, whereas at least a few data collection systems with which we are familiar place the time before the code. This is hardly problematic. In such cases, it is relatively easy to write a simple computer program (in Pascal, for example) that reformats the data as initially collected into SDIS format.

No matter the particular form – whether event, state, timed-event, or interval sequences – the advantages of a few standard forms for sequential data are considerable. They make it easier, and more worthwhile, to develop general-purpose programs for sequential data – such as GSEQ (Bakeman & Quera, 1995a) – that compute, not just simple frequencies and percentages for different codes, but a variety of conditional probabilities and sequential and other statistics as well. Such general-purpose programs can then be shared among different laboratories. In addition, relying on a standard form for data representation should enhance the development of special-purpose software within a given laboratory as well.

5.6 Cross-classified events

In chapter 3, we distinguished between two general approaches to data collection: intermittent versus continuous recording. Because this is a book about sequential analysis, we stressed continuous recording strategies, defining four particular ones: (a) coding events, (b) recording onset and offset times, (c) timing pattern changes, and (d) coding intervals. We also said that a fifth strategy – cross-classifying events – could result in sequential data if the major categories used to cross-classify the event could be arranged in a clear temporal order.

When continuous recording strategies are used, there is some choice as to how data should be represented. When events are cross-classified, however, there seems only one obvious way to do it: Each line represents an event, each column a major category. For example, as described in section 2.13, Bakeman and Brownlee coded object struggles. The first major category was prior possession, the second was resistance, and the third was success. Thus (if 1 = yes, 2 = no) the following

```
1  1  1
2  1  2
```

would code two events. In the first, the child attempting to take the object had had prior possession (1), his take attempt was resisted (1), but he succeeded in taking the object (1). In the second, the attempted taker had not had prior possession (2), he was likewise resisted (1), and in this case, the other child retained possession (2).

Unlike the SDIS data formats discussed in the previous several sections, data files that contain cross-classified event data are no different from the usual cases by variables rectangular data files analyzed by the standard statistical packages such as SPSS and SAS. Typically, cross-classified data are next subjected to log-linear analyses. When events have been detected and cross-classified in the first place, the data file could be passed, more or less unaltered, to the log-linear routines within SPSS or other standard packages, or to a program like ILOG (Bakeman & Robinson, 1994) designed specifically for log-linear analysis. Moreover, often analyses of SDIS data result in contingency table data, which likewise can be subjected to log-linear analysis with any of the standard log-linear programs. In fact, GSEQ is designed to examine sequential data and produce contingency table summaries in a highly flexible way; consequently it allows for the export of such data into files that can subsequently be read by SPSS, ILOG, or other programs.

Table 5.1. *Relationship between data recording and data representation*

This data recording strategy	Allows for this data representation
Coding events, no time	Events sequences
Recording onset and offset times, or timing pattern changes, or coding intervals	Events, state, timed-event, or interval sequences
Cross-classifying events	Cross-classified events

5.7 Transforming representations

Early in this chapter, we suggested that the data as collected should not become a straitjacket for subsequent analyses; that it was important to put the data into a form convenient for analysis. A corollary is that the data as collected may take various forms for different analyses and that one form may be transformed into another. There are limits on possible transmutations, of course. Silken detail cannot be extracted from sow-ear coarseness. Still, especially when onset and offset times have been recorded, the data as collected can be treated as a data "gold mine" from which pieces in various forms can be extracted, tailored to particular analyses.

The relationship between data recording strategies and data representation form is presented in Table 5.1. As can be seen, when onset and offset or pattern-change times are recorded, aspects of that collected data can be represented as event, state, timed-event, or interval sequences. Further data represented initially as state or timed-event sequences can be transformed into event or even interval sequences; again, see Bakeman and Quera (1995a) for details.

One example of the potential usefulness of data transformation is provided by Bakeman and Brown's (1977) study of early mother–infant interaction. Desiring to establish an "ethogram" of early interactive behavior in the context of infant feeding, they defined an extensive number of detailed behavioral codes, more than 40 for the infant and 60 for the mother. Some of these were duration events, some momentary, the whole represented as timed-event sequences. For one series of analyses, Bakeman and Brown wanted to examine the usefulness of viewing interaction as a "behavioral dialogue." To this end, they defined some of the mother codes and some of the infant codes as representing "communicative acts," actions that seemed potentially communicative or important to the partner.

This done, they then proceeded to extract interval sequences from the timed-event sequential data. Each successive interval (they used a 5-second interval, but any "width" interval could have been used) was categorized as follows: (a) interval contains neither mother nor infant "communicative act" codes, (b) interval contains some mother but no infant codes, (c) interval contains some infant but no mother codes, and (d) interval contains both mother and infant codes. This scheme was inspired by others who had investigated adult talk (Jaffe & Feldstein, 1970) and infant vocalization and gaze (Stern, 1974). With it, Bakeman and Brown were able to show, in a subsequent study, differences between mothers interacting with preterm and full-term infants (Brown & Bakeman, 1980). But the point for now is to raise the possibility, and suggest the usefulness, of extracting more than one representation from the data as originally recorded.

A second example of data transformation is provided by the Bakeman and Brownlee (1980) study of parallel play described in chapter 1. There an interval recording strategy was used. Each successive 15-second interval was coded for predominant play state as follows: (a) Unoccupied, (b) Solitary, (c) Together, (d) Parallel, or (e) Group play. Thus the data as collected were already in interval sequential form and were analyzed in this form in order to determine percentage of intervals assigned to the different play states.

However, to determine if these play states were sequenced in any systematic way, Bakeman and Brownlee transformed the interval sequence data into event sequences, arguing that they were concerned with which play states followed other states, not with how long the preceding or following states lasted. But once again, the moral is that for different questions, different representations of the data are appropriate.

Throughout this book, the emphasis has been on nominal-scale measurement or categorization. Our usual assumption is that some entity–an event or a time interval – is assigned some code defined by the investigator's coding scheme. But quantitative measurement can also be useful, although such data call for analytic techniques not discussed here. (Such techniques are discussed in Gottman, 1981.) The purpose of this final example of data transformation is to show how categorical data can be transformed into quantitative time-series data.

Tronick, Brazelton, and their co-workers have been interested in the rhythmic and apparently reciprocal way in which periods of attention and nonattention, of quiet and excitation, seem to mesh and merge with each other in the face-to-face interaction of mothers with their young infants (e.g., Tronick, Als, & Brazelton, 1977). They videotaped mothers and infants interacting and then subjected those tapes to painstaking coding, using an interval coding strategy. Several major categories were defined, each containing a number of different codes. The major categories included, for example, vocalizations, facial expressions, gaze directions, and body movement for both mother and infant. The tapes were viewed repeatedly, often in slow motion. After each second of real time, the observers would decide on the appropriate code for each of the major categories. The end result was interval sequential data, with each interval representing 1 second and each containing a specific code for each major category.

Next, each code within each of the major categories was assigned a number, or weight, reflecting the amount of involvement (negative or positive) Tronick thought that code represented. In effect, the codes within each major category were ordered and scaled. Then, the weights for each category were summed for each second. This was done separately for mother and infant codes so that the final result was two parallel strings of numbers, or two times series, in which each number represented either the mother's or infant's degree of involvement for that second. Now analyzing two time series for mutual influence is a fairly classic problem, more so in astronomy and economics than psychology, but transforming observational data in this way allowed Gottman and Ringland (1981) to test directly and quantitatively the notion that mother and infant were mutually influencing each other.

5.8 Summary

Five standard forms for representing observational data are presented here. The first, event sequences, consists simply of codes for the events, ordered as they occurred. The second, state sequences, adds onset times so that information such as proportions of time devoted to different codes and average bout durations can be computed. The third, timed-event sequences, allows for events to cooccur and is more open-ended; momentary and duration behaviors are indicated along with their onset and offset times, as required. The fourth, interval sequences, provides a convenient way to represent interval recorded data. And the fifth form is for cross-classified events.

An important point to keep in mind is that data as collected can be represented in various ways, depending on the needs of a particular analysis. Several examples of this were presented in the last section. The final example, in fact, suggested a sixth data representation form: time series. Ways to analyze all six forms of data are discussed in the next chapters, although the emphasis is on the first five. (For time-series analyses, see Gottman, 1981.) Some of the analyses can be done by hand, but most are facilitated by using computers. An advantage of casting data into these standard forms is that such standardization facilitates the development and sharing of computer software to do the sorts of sequential analyses described throughout the rest of this book. Indeed, GSEQ (Bakeman & Quera, 1995a) was developed to analyze sequential data represented according to SDIS conventions.

6

Analyzing sequential data:
First steps

6.1 Describing versus modeling

If all the steps described in previous chapters – developing coding schemes, recording behavioral sequences reliably, representing the observational data – are in order, then the first fruits of the research should be simple description. Introductory textbooks never tire of telling their readers that the basic tasks of psychology are, one, description, and two, explanation. Similarly, almost all introductory textbooks in statistics distinguish between descriptive statistics, on the one hand, and inferential statistics, on the other. This distinction is important and organizes not just introductory statistics texts but this and the next four chapters as well.

Much of the material presented in this and the following four chapters, however, assumes that readers want first to describe their data, and so description is emphasized. Problems of inference and modeling – determining if data fit a particular model, estimating model parameters – are touched on only slightly here. These are important statistical topics and become especially so when one wants to move beyond mere description to a deeper understanding of one's data. That is why so many books and courses, indeed huge specialized literatures, are devoted to such topics. We assume that readers will use scores derived from observing behavioral sequences as input for anything from simple chi-square or analyses of variance, to log-linear modeling, to the modeling approach embodied in programs like LISREL.

Our task, fortunately, is not to describe all the modeling possibilities available. Instead, we have set ourselves the more manageable task of discussing how to derive useful descriptive scores from sequential data. Still, we describe some simple instances of model testing and try to point out when sequential data analysis presents particular problems for statistical inference. Throughout, we attempt to maintain the distinction between description and modeling. Thus we would never talk of "doing" a Markov analysis. Instead we would describe how to compute transitional probabilities, on the one hand, and how to determine if those transitional probabilities fit a particular Markov model, on the other (see section 8.3).

In the remainder of this chapter, we note some of the simpler descriptive statistics that can be derived from sequential observational data.

6.2 Rates and frequencies

Some statistics are so obvious that it almost seems an insult to the reader's intelligence to mention them at all. Still, they are so basic, and so useful, that omitting them would seem negligent. Certainly, logical completeness demands their inclusion. One such statistic is the rate or frequency with which a particular event occurred.

What is recorded is the *event*. This could be either a momentary or a duration event. Just the occurrence of the event might be recorded, or its onset and offset times might be recorded as well, or the occurrence of events might be embedded in another recording strategy such as coding intervals or cross-classifying events.

What is tallied is the *frequency* – how often a particular event occurred. The data could be represented as simple event sequences, as state sequences, as timed-event sequences, or as cross-classified events. In all these cases, it is possible to count how often particular events occurred. Only interval sequences pose potential problems. Two successive intervals that contain the same code may or may not indicate a single bout; thus frequencies of intervals that contain a particular code may overestimate the number of instances, and this needs to be taken into account when interpreting frequency data for interval sequences.

In most cases, raw frequencies should be transformed to *rates*. frequencies, of course, depend on how long observation continues, whereas rates have the merit of being comparable across cases (individuals, days, etc.). The total observation time needs to be recorded, of course, even for event sequences, which otherwise do not record time information; otherwise rates cannot be computed. For example, Adamson and Bakeman (1985) observed infants at different ages (6, 9, 12, 15, and 18 months) and with different partners (with mothers, with peers, and alone). A coder recorded whenever infants engaged in an "affective display." For a variety of reasons, the observation sessions (with each partner, at each age) varied somewhat in length. Thus Adamson and Bakeman divided the number of affective displays (for each observation session) by the observation time (in their case, some fraction of an hour), yielding a rate per hour for affective displays. This statistic, just by itself, has descriptive value (there were 59 affective displays per hour, on the average), but further, it can be (and was) used as a score for traditional analysis of variance. Thus Adamson and Bakeman were able to report that the rate of affective displays was greater when infants were with mothers instead of peers, and that this rate increased with age.

6.3 Probabilities and percentages

A second obvious and very useful descriptive statistic is the simple probability or percentage. It can be either event-based or time-based. When event-based, the simple probability or percentage tells us what proportion of events were coded in a particular way, relative to the total number of events coded. Initial procedures are the same as for rates. Events are recorded (using any of the recording strategies described in chapter 3) and the frequencies tallied. The only assumption required is that codes be mutually exclusive and exhaustive. Then the number of events coded in a particular way is divided by the total number of events coded. This gives the simple probability for that particular kind of event. Or, the quotient can be multiplied by 100, which gives the percentage for that particular kind of event. For example, in the Bakeman and Brownlee study of parallel play, 12% of the events were coded Unoccupied, 18% Solitary, 26% Together, 24% Parallel, and 21% Group. (Recall that they used an interval-coding strategy; thus a single event was defined as any contiguous intervals coded the same way.)

When simple probabilities or percentages are time based, the interpretation is somewhat different. These widely used statistics convey "time-budget" information, that is, they indicate how the cases observed (animals, infants, children, dyads, etc.) spent their time. The recording strategy must preserve time information; thus simple event coding would not work, but recording onset and offset times of events, or even coding intervals (remembering the approximate nature of the statistics estimated) would be fine. Similarly, data would need to be represented as state, timed-event, or interval sequences, not simple event-sequences. (Cross-classified event data would also work if duration had been recorded, but frequently when this recording and representation approach is used, the proportion of time devoted to the event being cross-classified may not be recorded.) One final note: When proportion of time coded in a particular way is of interest, codes need not be mutually exclusive and exhaustive.

The Bakeman and Brownlee study of parallel play, just used as an example of event-based probabilities, also provides an example of time-based probabilities, or percentages as well. Using an interval-coding strategy, they reported that 9% of the 15-second intervals were coded Unoccupied,

25% Solitary, 21% Together, 28% Parallel, and 17% Group. Thus, for example, although 24% of the events were coded Parallel play, Parallel play occupied 28% of the time. (These are estimates, of course. Recording onset times for behavioral state changes would have resulted in more accurate time-budget information than the interval-recording strategy actually used; see section 3.7.) A second example could be provided by the Adamson and Bakeman study of affective displays. They recorded not just the occurrence of affective displays, but their onset and offset times as well. Thus we were able to compute that affective displays occurred, on the average, 4.4% of the time during observation sessions. Put another way, the probability that the infant would display affect in any given moment was .044.

Event-based (e.g., proportion of events coded Solitary) and time-based (e.g., proportion of time coded Solitary) probabilities or percentages provide different and independent information; there is no necessary correlation between the two. Which then should be reported? The answer is, it depends. Whether one or both are reported, investigators should always defend their choice, justifying the statistics reported in terms of the research questions posed.

6.4 Mean event durations

Whenever time information is recorded, mean event durations can be reported as well as (or instead of) proportions of total time devoted to particular kinds of events. In fact, mean event durations provide no new information not already implied by the combination of rates and time-based percentages. After all, mean event (or bout, or episode) durations are computed by dividing the amount of time coded for a particular kind of event by the number of times that event was coded. But in some cases, mean event durations may be more useful descriptively than time-based probabilities or percentages.

Because of the clear redundancy among these three descriptive statistics (rates or frequencies, time-based probabilities or percentages, mean event durations), we think investigators should report, or at least analyze, only two of them. The question then is, which two? The answer will depend on whatever an investigator thinks most useful descriptively, given the sort of behavior coded. However, we suspect that when behavioral states are being coded (see section 3.2), time-based percentages and mean durations are more useful than rates, but that when the events being coded occur just now and then and do not exhaustively segment the stream of behavior as behavioral states do, then rates and mean event durations may prove more useful than time-based percentages.

For example, although in the previous section we computed the (time-based) probability of an affective display from the Adamson and Bakeman data, they in fact repor~~ted~~ ~~~~s for affective displays. As with rates, m~~~~ore for subsequent analysis of variance. T~~~~orted, not just that the average length of ~~~~onds, but that the length became shorter ~~~~ This was an effect Adamson and Bakema~~n~~ ~~~~ hypothesized shift in the function of affect~~~~ages.

6.5 Transition~~al probabilities: An introducti~~on

The statistics discussed ~~~~ preceding sections can be extremely useful for describing aspects of sequential observational data (and all can be computed with the GSEQ program), but they do not themselves convey anything uniquely sequential. Perhaps the simplest descriptive statistic that does capture a sequential aspect of such data is the transitional probability; however before transitional probabilities can be described, some definitions are in order.

A simple (or unconditional) probability is just the probability with which a particular "target" event occurred, relative to a total set of events (or intervals, if time based). For example, if there were 20 days with thunderstorms last year, we could say that the probability of a thunderstorm occurring on a particular day was .055 (or 20 divided by 365).

A conditional probability, on the other hand, is the probability with which a particular "target" event occurred, relative to another "given" event. Thus if it rained 76 days last year, and if on 20 of those 76 days there were thunderstorms, then we would say that the probability of a thunderstorm occurring, given that it was a rainy day, was .263 (or 20 divided by 76). If T stands for thunderstorms and R for a rainy day, then the simple probability for thunderstorms is usually written $p(T)$, whereas the conditional probability for thunderstorms, given a rainy day, is usually written $p(T|R)$; in words, this is "the probability of T, given R."

A transitional probability is simply one kind of conditional probability. It is distinguished from other conditional probabilities in that the target and given events occur at different times. Often the word "lag" is used to indicate this displacement in time. For example, if data are represented as event sequences, then we might want to describe the probability, given event A, of the target event B occurring immediately after (lag 1), occurring after an intervening event (lag 2), etc. These event-based transitional probabilities can be written $p(B_{+1}|A_0)$, $p(B_{+2}|A_0)$, etc.

Similarly, if data are represented as state, timed-event, or interval sequences, then we might want to describe the probability, given event A, of the target event B occurring in the next interval (often written $t + 1$, where t stands for time, in the interval after the next ($t + 2$), etc. Such time-based probabilities are often written $p(B_{t+1}|A_t)$, $p(B_{t+2}|A_t)$, etc.

For example, consider the following sequence. Assume that each letter stands for an event and that this sequence contains three mutually exclusive and exhaustive codes.

    $B \ C \ A \ B \ B \ C \ B \ C \ A \ C$

In this case, it turns out that each code occurred four times. That is, $f(A) = f(B) = f(C) = 4$, therefore $p(A) = p(B) = p(C) = 4/12 = .33$, because 12 events were coded in all. The transitional frequency matrix for these data is given in Figure 6.1. Note that the labeling of rows and columns is somewhat arbitrary. We could just as well have labeled rows lag-1 and columns lag 0. Either way, rows refer to events occurring earlier, columns to events occurring later in time. This is the usual convention (probably because we are a left-to-right reading society) and one we recommend following.

Transitional frequency matrices are easy to construct. Each cell indicates the number of times a particular transition occurred. For example, B was followed by C three times in the sequence given in the preceding paragraph; thus the cell formed by the Bth row and the Cth column in the lag-one frequency matrix contains a "3." Symbolically, $f(C_{+1}|B_0) = 3$, or $f_{BC} = 3$; this if often written x_{BC}, letting x represent a score in general or, less mnemonically but more conventionally, x_{23}. Note that if N successive events or intervals are coded, then there will be $N - 1$ lag-one transitions, $N - 2$ lag-two transitions, etc. (The reader may want to verify the other frequencies for the transitional frequency matrices given in Figure 6.1.)

To tally transitions, we use a "moving time-window." For the lag 1 transition matrix, we slide the two-event moving time-window along as follows:

    $(B \ C) \ A \ A \ A \ B \ B \ C \ B \ C \ A \ C$

Then,

    $B \ (C \ A) A \ A \ B \ B \ C \ B \ C \ A \ C$

Then,

    $B \ C \ (A \ A) \ A \ B \ B \ C \ B \ C \ A \ C$

etc. The first position adds a tally to the x_{BC} cell of the first table in Figure 6.1. The second position of the window adds a tally to the x_{CA} cell, and so forth. Note that the consequent code takes its turn next as an antecedent. This raises questions of independence of tallies, and may

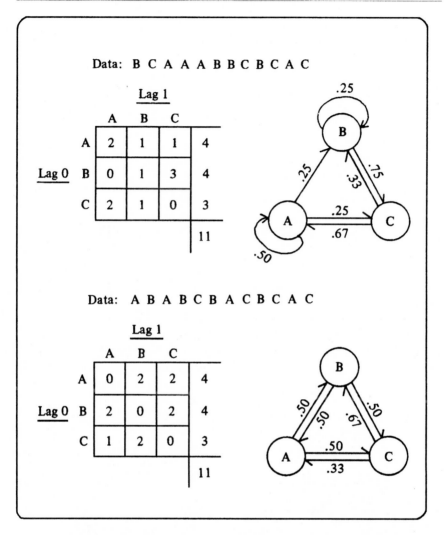

Figure 6.1. Examples of transitional frequency matrices and state transition diagrams.

matter for some subsequent inferential statistical procedures; this issue is discussed further in section 8.1, although Bakeman and Dorval (1989) believe it is not practically consequential.

Similarly, transitional probability matrices are easy to compute. For example, for the data sequence given earlier, the lag 1 transitional probability matrix would be as given in Table 6.1. As the reader can see, a

Table 6.1. *Transitional probability matrix for*
the first data sequence given in Figure 6.1

	Lag 1		
Lag 0	A	B	C
A	.50	.25	.25
B	.00	.25	.75
C	.67	.33	.00

transitional probability is just the frequency for a particular cell divided by
the frequency for that row. (A consequence of this definition is that the
transitional probabilities in each row sum to 1.) Symbolically,

$$t_{ij} = \frac{x_{ij}}{x_{i+}}$$

For example, the probability of code C occurring, given that code B just
occurred, is $p(C_{+1}|B_0) = t_{BC} = x_{BC} \div x_{B+} = 3/4 = .75$. This means
that 75% of the time, code C followed code B.

Transitional probabilities are often presented graphically, as state transition diagrams (for examples, see Bakeman & Brown, 1977; Stern, 1974).
Such diagrams have the merit of rendering quite visible just how events (or
time intervals) were sequenced in time. Circles represent the codes, and
arrows represent the transitional probabilities among them. Examples are
given in Figure 6.1, and the reader may want to verify that they were drawn
and labeled correctly.

Figure 6.1 contains a second data sequence, along with its associated
transitional frequency matrix and state transition diagram. For both, the
simple probabilities are the same, that is, for both sequences each different
code occurred four times. The point of presenting these two examples
is to show that, even when simple probabilities indicate no differences,
events may nonetheless be sequenced quite differently. And when they
are, transitional probabilities and their associated state transition diagrams
can reveal those differences in a clear and informative way.

One final point: The discussion here treats transitional probabilities as
simple descriptive statistics, and state transition diagrams as simple descriptive devices. Others (e.g., Kemeny, Snell, & Thompson, 1974) discuss transitional probabilities and state transition diagrams from a formal,
mathematical point of view, as parameters of models. Interesting questions
for them are, given certain models and model parameters, what sorts of outcomes would be generated? This is a formal, not an empirical exercise, one
in which data are generated, not collected. For the scientist, on the other

hand, the usual question is, first, can I accurately describe my data, and second, does a particular model I would like to support generate data that fit fairly closely with the data I actually got? The material in this chapter, as noted in its first section, is concerned mainly with the first enterprise – accurate description.

6.6 Summary

This "first steps" chapter discusses four very simple, very basic, but very useful, statistics for describing sequential observational data: rates (or frequencies), simple probabilities (or percentages), mean event durations, and transitional probabilities.

Rates indicate how often a particular event of interest occurred. They are probably most useful for relatively momentary and relatively infrequent events like the affective displays Adamson and Bakeman (1985) described. Simple probabilities indicate what proportion of all events were of a particular kind (event based) or what proportion of time was devoted to a particular kind of event (time based). Time-based simple probabilities are especially useful when the events being coded are conceptualized as behavioral states (see section 3.2). Indeed, describing how much time individuals (dyads, etc.) spent in various activities (or behavioral states) is a frequent goal of observational research.

With the knowledge of how often a particular kind of event occurred, and what percentage of time was devoted to it, it is possible to get a sense of how long episodes of the event lasted. Mean event durations can be computed directly, of course. But because these three statistics provide redundant information, the investigator should choose whichever two are most informative, given the behavior under investigation.

Finally, transitional probabilities capture sequential aspects of observational data in a simple and straightforward way. They can be presented graphically in state transition diagrams. Such diagrams have the merit of rendering visible just how events are sequenced in time.

7

Analyzing event sequences

7.1 Describing particular sequences: Basic methods

Throughout this chapter, we assume that the reader is interested in event-sequence data. That is, no matter how the data may have been recorded and represented initially, we assume that it is possible to extract event sequences (see section 5.2) from the data, and that the investigator has good reasons for wanting to do so. This means that the data to be analyzed are represented as sequences or chains of coded events (or behavioral states but without time information) and that those events are defined in a way that makes them mutually exclusive and exhaustive. Sometimes the chains will be unbroken, collected all during one uninterrupted observation session. Other times, several sequences may be pooled together for an analysis, either because there were breaks in the observation session or because observation sessions occurred at different times. Data from more than one subject may even be pooled for some analyses (see section 8.4). In all cases, the data to be analyzed consist of chains or sequences of codes.

The codes for event sequences are mutually exclusive and exhaustive, as already noted. In addition, often the logic of the situation does not permit consecutive codes to repeat. For example, when coders are asked to segment the stream of behavior into behavioral states, it follows naturally that two successive states cannot be coded the same say. If they were, they would not be two states, after all, but just one. The restriction that the same code may not follow itself in event sequences occurs relatively often, especially when the events being coded are thought of as states. This restriction affects how some statistics, especially expected frequencies, are computed, as we discuss later.

As an example, let us again assume the coding scheme used in the Bakeman and Brownlee study of parallel play. Behavioral states were classified as one of five kinds: Unoccupied, Solitary, Together, Parallel, and Group. When sequences were analyzed, adjacent codes were not allowed to be identical. Thus there were 20 (or 5 × 4) different kinds of two-event sequences (not 5^2, which would be the case if adjacent codes could be the

100

same); 80 (or 5×4^2) different kinds of three-event sequences (not 5^3); 320 (or 5×4^3) different kinds of four-event sequences (not 5^4); etc.

Determining how often particular two-event, three-event, etc., sequences occurred in one's data is what we mean by "basic methods." This involves nothing more than counting. The investigator simply defines particular sequences, or all possible sequences of some specified length, and then tallies how often they appear in the data. For example, Bakeman and Brownlee were particularly interested in transitions from Parallel to Group play. For one child in their study, 127 two-event sequences were observed, 10 of which were from Parallel to Group. Thus they could report that, for that child, f(PG) = 10 and p(PG) = .079 (10 divided by 127). (Note that p(PG) is not a transitional probability. It is the simple or zero-order probability for the two-event sequence, Parallel to Group.) In sum, the most basic thing to do with event-sequence data is to define particular sequences, count them, and then report frequencies and/or probabilities for those sequences.

7.2 Determining significance of particular chains

We might now ask, how should these values be evaluated? One possibility would be to compute expected frequencies, based on some model, for particular observed frequencies, and then compare observed and expected with a chi-square goodness-of-fit test. For example, there are 20 different kinds of two-event sequences possible (U to S, U to T, U to P, U to G, S to U, S to T, etc.). Thus, we might argue, the expected probability for any one kind is .05 (1/20), and so the expected frequency in this case is 6.35 (.05 × 127). What we are doing is assuming a particular model – in this case, a "zero order" or "equiprobable" model, so called because it assumes that the five codes occur with equal probability – and then comparing the expected values the model generates for a particular sequence with those actually observed. We note that the observed value for the Parallel to Group sequence, 10, is greater than the expected value, 6.35. If we were only concerned with the Parallel to Group sequence, we might categorize all sequences as either Parallel to Group (10) or not (117), and compare observed to expected using the familiar Pearson chi-square statistic,

$$X^2 = \sum \frac{(obs - exp)^2}{exp}$$
$$= \frac{(10 - 6.35)^2}{6.35} + \frac{(117 - 120.65)^2}{120.65} = 2.21$$

which, with one degree of freedom, is not significant (we use a Roman X to represent the computed chi-square statistic to distinguish it from a Greek chi, which represents the theoretical distribution).

Alternatively, we might make use of what we already know about how often the five different codes occurred. This "first-order" model assumes that codes occurred as often as they in fact did (and were not equiprobable), but that the way codes were ordered was determined randomly. For the child whose data we are examining, 143 behavioral states were coded; 34 were coded Parallel and 30 Group. (Because 127 two-event sequences were tallied, and 143 states were coded, there must have been 15 breaks in the sequence.) Now, if codes were indeed ordered randomly, then we would expect that the probability for the joint event of Parallel followed by Group would be equal to the simple probability for Parallel multiplied by the simple probability for Group (this is just basic probability theory). Symbolically,

$$p(PG)_{exp} = p(P) \times p(G)$$

The $p(P)$ is .238 (34/143, the frequency for Parallel divided by the total, N).

In this case, however, the $p(G)$ is not the $f(G)$ divided by N. Because a Parallel state cannot follow a Parallel state, the probability of group (following Parallel) is computed by dividing the frequency for Group, not by the total number of states coded, but by the number that could occur after Parallel – that is, the total number of states coded, less the number of Parallel codes. Symbolically.

$$p(G) = \frac{f(G)}{N - f(P)}$$

when adjacent codes must be different and when we are interested in the expected probability for Group following Parallel. Now we can compute the expected probability for the joint event of a Parallel to Group transition. It is:

$$p(PG)_{exp} = \frac{f(P)}{N} \times \frac{f(G)}{N - f(P)} = \frac{34}{143} \times \frac{30}{143 - 34} = .0654$$

The expected frequency, then, is 8.31 (.0654, the expected probability for this particular two-event sequence, times 127, the number of two-event sequences coded).

The chi-square statistic for this modified expected frequency is

$$X^2 = \frac{(10 - 8.31)^2}{8.31} + \frac{(117 - 118.69)^2}{118.69} = 0.368$$

which likewise is not statistically significant. Neither analysis suggests that Group is any more likely to follow Parallel than an equiprobable or first-order independence model would suggest.

The methods presented in this section are fairly limited. First, as sequences become longer, the number of possible sequences increases exponentially. For example, with just five codes when consecutive codes cannot repeat, there are 20 (5 × 4) two-event sequences, 80 (5 × 4 × 4) three-event sequences, 320 (5 × 4 × 4 × 4) four-event sequences, etc. Consequently, expected probabilities for any one sequence may become vanishingly small, requiring staggering amounts of data before expected frequencies become large enough to evaluate with any confidence. Second, rarely are investigators interested in just one particular sequence such as the Parallel to Group sequence used here as an example. More general methods, described in subsequent sections and chapters, are required.

7.3 Transitional probabilities revisited

In the last section, we presented data derived from observing one child in the Bakeman and Brownlee study of parallel play. We noted that her event-sequence data contained 127 two-event sequences and that 10 of them represented transitions from Parallel to Group play. Thus we were able to say that p(PG), the probability of a Parallel to Group sequence, was .0787, or 10 divided by 127. We then discussed ways of determining whether this observed probability differed significantly from expected. We also noted that p(PG) was a simple, not a transitional probability. In other words, p(PG) is the probability for this particular sequence; if the probabilities for all 20 possible two-event sequences were summed, they would add up to one.

Occasionally it may be useful to describe probabilities for particular sequences, no matter whether chains are two-event, three-event, or longer. When longer sequences are considered (e.g., 4 or 5 events long instead of just 2), the number of possible sequences increases exponentially. When there are many possible sequences, probabilities for particular sequences can become almost vanishingly small and, as a result, less useful descriptively. Thus usually attention focuses on transitional probabilities involving two events. As discussed in section 6.5, these are usually symbolized t and, unless noted otherwise, refer to lag 1.

For example, consider again the child in the parallel play study. (frequencies and simple and transitional probabilities derived from her data are given in Tables 7.1 through 7.3.) Considering just simple probabilities, we note that the probability of a Parallel to Group sequence, p(PG), was .0787, whereas the probability of a Parallel to Unoccupied sequence,

Table 7.1. *Observed frequencies for two-event sequences*

Given code,	Target code, lag 1					
lag 0	Un.	Sol.	Tog.	Par.	Gr.	Totals
Unoccupied	—	6	5	2	2	15
Solitary	5	—	6	7	5	23
Together	5	6	—	12	10	33
Parallel	2	7	11	—	10	30
Group	2	4	11	9	—	26
Totals	14	23	33	30	27	127

Table 7.2. *Simple probabilities for two-event sequences*

Given code,	Target code, lag 1				
lag 0	Un.	Sol.	Tog.	Par.	Gr.
Unoccupied	—	.0472	.0394	.0157	.0157
Solitary	.0394	—	.0472	.0551	.0394
Together	.0394	.0472	—	.0945	.0787
Parallel	.0157	.0551	.0866	—	.0787
Group	.0157	.0315	.0866	.0709	—

Note: The tabled probabilities do not sum exactly to 1 because of rounding.

p(PU), was .0157. This certainly conveys the information that Group was more common after Parallel than Unoccupied. But somehow it seems both clearer and descriptively more informative to say that the probability of Group, given a previous Parallel, p($G|P$) or t_{PG}, was .333, whereas the probability of Unoccupied, given a previous Parallel, p($U|P$) or t_{PU}, was .067. Immediately we know that 33.3% of the events after Parallel were Group, whereas only 6.7% were Unoccupied.

We just considered transitions from the same behavioral state (Parallel) to different successor states (Group and Unoccupied), but the descriptive value of transitional probabilities is portrayed even more dramatically when transitions from different behavior states to the same successor state are compared. For example, the simple probabilities for the Unoccupied to Solitary, p(US), and for the Together to Solitary, p(TS), transitions are both .0472. Yet the probability of Solitary, given a previous Unoccupied, p($S|U$) or t_{US}, is .400, whereas the probability of Solitary, given a previous Together, p($S|T$) or t_{TS}, is .182. The transitional probabilities "correct"

Table 7.3. *Transitional probabilities for two-event sequences*

Given code,	Target code, lag 1			
lag 0		Tog.	Par.	Gr.
Unoccu...		.333	.133	.133
Solitary		.261	.304	.217
Togeth...		—	.364	.303
Parallel		.367	—	.333
Group		.423	.346	—

Note: ...ding.

for difference ... behavioral states and, therefore, clearly r... was relatively common after Unoccupied, and considerably less so after Together, even though the Unoccupied to Solitary and the Together to Solitary transitions appeared the same number of times in the data.

Moreover, as already noted in section 6.5, transitional probabilities form the basis for state transition diagrams, which, at least on the descriptive level, are a particularly clear and graphic way to summarize sequential information. The only problem is that, even with as few as five states, the number of possible arrows in the diagram can produce far more confusion than clarity. The solution is to limit the number of transitions depicted in some way. In this case, for example, we could decide to depict only transitional probabilities that are .3 or greater, which is what we have done in Figure 7.1. This reduces the number of arrows in the diagram from a possible 20, if all transitions were depicted, to a more manageable 9.

The nine transitions shown in Figure 7.1 are not necessarily the most frequent transitions; this information is provided by the simple probabilities for two-event sequences (see Table 7.2). Nor are the transitions necessarily significantly different from expected; to determine this we would need to compute and evaluate a z score for each transition (see next section). What the state transition diagram does show are the most likely transitions, taking the base rate for previous states into account. In other words, it shows the most likely ways of "moving" from one state to another. For this one child, Figure 7.1 suggests frequent movement from Unoccupied to both Solitary and Together, from Solitary to Parallel, and reciprocal movement among Together, Parallel, and Group.

One final point: Transitional probabilities can be used to describe relationships between two nonadjacent events as well. Not only can we compute, for example, the probability of Group in the lag 1 positions given an immediately previous Parallel: $p(G_{+1}|P_0)$, but we can also compute the

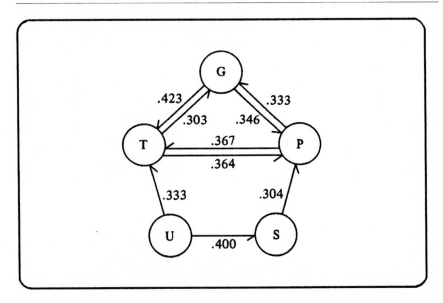

Figure 7.1. A state transition diagram. Only transitional probabilities greater than 0.3 are shown. U = unoccupied, S = solitary, T = together, P = parallel, and G = group.

probability of Group in the lag 2 position: $p(G_{+2}|P_0)$, the lag 3 position, etc., with as many events intervening between the "given" and the "target" code as desired. The ability of transitional probabilities to describe such lagged relationships, in fact, forms the basis for the lag sequential method described in section 7.5.

Transitional probabilities, although useful descriptively, also have limitations. In particular, when several "subjects" take part in a study, and when transitional probabilities are computed separately for each subject, those transitional probabilities should not, under most circumstances, be used as scores for testing for individual or group differences.

The reason is as follows: A transitional probability is valuable descriptively, to be sure, and certainly reflects the "experience" of a particular subject. For example, the experience of the child whose data are given in Table 7.3 was that one-third of the time after Parallel, Group followed. This may be important, but it may or may not be "significant." It depends on how probable Group was for the child. If Group was especially probable for one child, then a score of .33 for the transitional probability might not be very high; it could even be less than expected. On the other hand, if Group was unlikely for another child, then a score of .33 might be quite high, considerably above expected.

The major problem, then, with using transitional probabilities when analyzing for individual or group differences, is that similar numeric values may have quite different meanings for different subjects, or for the same subject at different times, rendering the results difficult to interpret at best. A secondary problem is that the values of transitional probabilities are "contaminated" with the values for the simple probabilities, which means that what appear to be analyses of transitional probabilities may in fact reflect little more than simple probability effects.

An example may help to clarify this. Imagine that we have two kinds of children, farm kids and city kids, and that the mean value for the simple probability of Group computed for the city kids is significantly higher than the mean value computed for farm kids. If we then turn around and test the transitional probability, $p(G_{+1}|P_0)$, we would expect that mean values for city and farm kids would differ significantly as well. After all, the expected value for this transitional probability is directly related to the simple probability for Group. The more often the Group code appears in the data, the more often we would expect it to follow the Parallel code as well.

Symbolically, what we are saying is that the expected value for $p(T|G)$ – where T stands for "target" and G for "given" – is directly related to $p(T)$, the probability of the target code. It is not necessary, of course, that observed agree with their expected values. Still, when analyses of both $p(T|G)$ and $p(T)$ reveal significant group differences, it seems unjustified to us to claim that the differences with respect to $p(T|G)$ are explained by anything more complex than the differences detected for $p(T)$. In sum, for both reasons given above, analyses that use transitional probabilities as scores rarely provide much insight into sequential aspects of the data.

Almost always, when individual or group differences are at issue, the appropriate score to use is not the transitional probability, but some index of the strength of the effect, like Yule's Q (as discussed in section 7.7). Analyzing such scores, we would be able to determine, for example, whether the Parallel to Group sequence was significantly more characteristic of city kids, on the average, than of farm kids, or whether the extent to which Group tended to follow Parallel was a stable characteristic of children measured at two different ages.

In the previous edition of this book, we suggested that z scores (see next section) be used as an individual index for such analyses, but the magnitude of z scores is affected by the number of tallies. For an effect of a specific size, the z score becomes larger as the number of tallies increases. This is an example of the well-known fact that power increases with sample size. But it also makes the z score an inappropriate choice for analyzing individual differences. Unless the number of tallies is the same for all subjects (dyads, etc.) analyzed, z scores conflate how strongly a particular transitional probability deviates from its expected value with the number of tallies available for the analysis.

7.4 Computing z scores and testing significance

Once lagged events have been tallied in a table like Table 7.1, where rows represent lag 0 and columns lag 1, and after we have examined transitional probabilities like those in Table 7.3, next we usually want to identify which transitional probabilities deviate significantly from their expected values. Commonly, a z score of some sort has been used for this purpose. Assuming that a computed score is distributed normally, then z scores larger than 1.96 absolute are often regarded as statistically significant at the .05 level. But there are a variety of ways to compute z scores, and simply calling one a z score hardly guarantees that it will be normally distributed.

Assume that the behavioral event of interest is called the "target" event, T, and that we want to relate that to another event, called the "given" event, G. In other words, we are interested in $p(T_1 | G_0)$ or t_{GT}, the probability of the target even occurring after the given event.

A z score compares observed to expected, so the first task is to compute the expected value for x_{GT}, the observed value for the transition from given to target behavior. When consecutive codes may repeat (and so the upper-left to lower-right diagonal tallies would not all be zero as in Table 7.1), expected values are computed using the familiar formula,

$$m_{GT} = \frac{x_{+T}}{x_{++}} x_{G+} = \frac{x_{G+} x_{+T}}{x_{++}} \tag{7.1}$$

where m_{GT} is an estimate of the expected frequency (m because often expected values are means), x_{G+} is the sum of the observed frequencies in the Gth or given row, x_{+T} is the sum of the observed frequencies in the Tth or target column, and x_{++} is the total number of tallies in the table (also symbolized as N_2 or the number of two-event chains tallied, as compared to N_1, the number of single events coded). This formula yields expected values assuming independence, that is, no association between the rows and columns of the table.

However, when consecutive codes cannot repeat, resulting in what are called structural zeros on the diagonal (*structural* because logical definition and not data collection resulted in their being zero), expected frequencies cannot be computed with a simple formula but require an iterative procedure, best performed with a computer. Two of the most widely used are iterative proportional fitting (IPF, also called the Deming–Stephan algorithm)

Table 7.4. *Expected frequencies for two-event sequences*

Given code,	Target code, lag 1				
lag 0	Un.	Sol.	Tog.	Par.	Gr.
Unoccupied	—	2.82	4.69	4.04	3.45
Solitary	2.65	—	7.83	6.75	5.76
Together	4.40	7.83	—	11.21	9.56
Parallel	3.80	6.75	11.21	—	8.24
Group	3.14	5.59	9.27	8.00	—

and the Newton–Raphson algorithm; for descriptions see Bishop, Fienberg, & Holland (1975) and Fienberg (1980). Both methods yields identical values. Expected frequencies for the data shown in Table 7.1 are given in Table 7.4.

When consecutive codes may repeat, z scores are computed as follows:

$$z_{GT} = \frac{x_{GT} - m_{GT}}{\sqrt{m_{GT}(1 - p_{G+})(1 - p_{+T})}} \tag{7.2}$$

where p_{G+} is $x_{G+} \div x_{++}$ and p_{+T} is $x_{+T} \div x_{++}$. In the log-linear literature, this is called an adjusted residual (Haberman, 1978, p.111). When consecutive codes cannot repeat, matters are more complex. The formula (actually several formulas, many of which are used by the Newton–Raphson algorithm) is given in Haberman (1979, p. 454; but see footnote 1 in Bakeman & Quera, 1995b); a number of relatively complex matrix operations are involved. The most practical way for you to compute these values, given structural zeros, is to use a computer program like GSEQ or a general-purpose log-linear program. Adjusted residuals, computed with the SPSS for Windows General Log-Linear routine, are given in Table 7.5.

If you are new to sequential analysis, you may want to skip the following paragraphs, which are included largely for historical purposes and for readers who wish to reconcile the preceding paragraphs with the first edition of this book and with earlier literature. Early on, Sackett (1979) suggested that z be computed as follows:

$$z_{GT} = \frac{x_{GT} - m_{GT}}{\sqrt{m_{GT}(1 - p_T)}} \tag{7.3}$$

where

$$m_{GT} = \mathrm{p}(T) \times \mathrm{f}(G) = \frac{x_T}{N_1} x_G = \frac{x_G x_T}{N_1} \tag{7.4}$$

which is almost but not quite the same as Equation 7.1 because it is based

Table 7.5. *Adjusted residuals for two-event sequences*

Given code,	Target code, lag 1				
lag 0	Un.	Sol.	Tog.	Par.	Gr.
Unoccupied	—	2.25	0.19	−1.29	−0.96
Solitary	1.71	—	−0.94	0.13	−0.42
Together	0.37	−0.94	—	0.37	0.22
Parallel	−1.17	0.13	−0.10	—	0.88
Group	−0.79	−0.88	0.85	0.51	—

Note: Row and column totals may not add exactly to those shown in Table 7.1 because of rounding.

on simple frequencies and probabilities for given and target behaviors, not on values from two-dimensional tables as is Equation 7.1. Equation 7.3 is based on the normal approximation for the binomial test,

$$z = \frac{x - NP}{\sqrt{NPQ}} \tag{7.5}$$

where N is f(G) or x_G, P is p(T) or $x_T \div N_1$ (also symbolized as p_T), and Q is $1 - p_T$. Almost immediately, however, Allison and Liker (1982) objected, noting that Equation 7.5 would only be appropriate if p_T were derived theoretically instead of from the data at hand. They wrote that Equation 7.3 should be

$$z_{GT} = \frac{x_{GT} - m_{GT}}{\sqrt{m_{GT}(1 - p_G)(1 - p_T)}} \tag{7.6}$$

instead, which is almost but not quite the same as Equation 7.2 because, like Equation 7.3, it is based on single occurrences and not two-event chains.

We prefer Equation 7.2 to 7.6 because it seems more grounded in a well-developed statistical literature, that dealing with log-linear models (e.g., see Bishop, Fienberg, & Holland, 1975; Fienberg, 1980; Wickens, 1989), and because it is based on two-event chains, which seems more faithful to the situation at hand. True, if a sequence of N_1 consecutive events is tallied using overlapped sampling (i.e., tallying first the e_1e_2 chain, then e_2e_3, e_3e_4, and so forth, where e stands for an event), so that $N_1 - 1$ chains are tallied, then x_G and x_{G+}, for example, will differ by at most 1. But overlapped sampling, while common, is not always used; moreover, often breaks occur in sequences and then the number of two-event chains tallied is $N_1 - S$, where S is the number of separate segments coded. In such cases, x_G and x_{G+} could differ by quite a bit.

Moreover, the log-linear tradition, from which Equation 7.2 is derived, offers a statistically based solution when consecutive codes cannot repeat (see Bakeman & Quera, 1995b). Sackett (1979), recognizing that Equation 7.1 would not compute expected frequencies correctly when structural zeros occupied the diagonal, suggested

$$m_{GT} = \frac{x_T}{N_1 - x_G}x_G = \frac{x_G x_T}{N_1 - x_G}. \tag{7.7}$$

He reasoned that when consecutive events cannot repeat, the expected probability for the target code at lag 1 (assuming independence) is the frequency for that code divided by the number of events that may occur at lag 1, which is N_1 minus the frequency for the given code. Thus expected frequencies on the diagonal are set to zero and off-diagonal ones are the frequency for the given code times this probability, as indicated by Equation 7.7. However, expected frequencies, when summed across rows and down columns, should equal the observed row and column totals, which expected frequencies computed per Equation 7.7 do not (Bakeman & Quera, 1995b), whereas expected frequencies computed with an iterative procedure do (see Table 7.4). Thus, as mentioned earlier, when consecutive codes cannot repeat, we would compute adjusted residuals using log-linear methods (and an appropriate computer program), not Equations 7.6 and 7.7.

7.5 Classic lag sequential methods

So far in this chapter, much of our discussion and most of our examples have been confined to two-event sequences. We have mentioned how longer sequences can be described and tested for significance (mainly in sections 7.1 and 7.2), but we have also noted that such tests may require prohibitive amounts of data. The approaches already discussed are "absolute" in the sense that they define particular sequences and then tally how often each occurred. As we attempt to investigate longer and longer sequences, the expected frequencies for particular sequences become vanishingly small, the number of possible sequences increases at a staggering rate, and it becomes almost impossible to make sense out of the wealth of information produced about so many different sequences. Clearly, a less absolute, more probabilistic and more flexible approach to the investigation of sequences comprising more than two events would be useful. One such approach is usually called the "lag sequential method." It was first developed by Sackett (1974, 1979, 1980) and later described by others (Bakeman & Dabbs, 1976; Bakeman, 1978; Gottman & Bakeman, 1979).

The reader of this book will already be familiar with the basic elements of the lag sequential method. As an example, assume that our code catalog defines several events, five of which are:

1. Infant Active
2. Mother Touch
3. Mother Nurse
4. Mother Groom
5. Infant Explore

(These codes are suggested by Sackett's work with macaque monkeys. The example here is based on one given in Sackett, 1974.) Assume further that successive events have been coded so that, as throughout this chapter, we are analyzing event-sequence data. Finally, assume that we are particularly interested in what happens after times when the infant is active, that is, we want to know whether there is anything systematic about the sequencing of events beginning with Infant Active episodes.

To begin with, the investigator selects one code to serve as the "criterion" or "given" event. In this case, that code would be Infant Active. Next, another code is selected as the "target." For example, we might select Mother Touch as our first target code. Then, a series of transitional probabilities are computed: for the target immediately after the criterion (lag 1), after one intervening event (lag 2), after two intervening events (lag 3), etc. Symbolically, we would write these lagged transitional probabilities as $p(T_1|G_0)$, $p(T_2|G_0)$, $p(T_3|G_0)$, etc. (remember, if we just write $p(T|G)$, target at lag 1 and given at lag 0 are assumed). The result is a series of transitional probabilities, each of which can then be tested for significance. For example, given Infant Active at lag "position" 0, if we had computed transitional probabilities for Mother Touch at lags 1 through 6, but only the lag 1 transitional probability significantly exceeded its expected value, we would conclude that Mother Touch was likely to occur just after Infant Active, but was not especially likely in the other lag "positions" investigated.

If we stopped now, we would have examined transitional probabilities for one particular target code, at different lags after a particular criterion code. This is not likely to tell us much about multievent sequences. The next step is to compute other series of transitional probabilities (and determine their significance), using the same criterion code but selecting different target codes. For example, given a criterion of Infant Active at lag 0, we could compute lag 1 through 6 transitional probabilities for Mother Nurse, Mother Groom, and Infant Explore. Imagine that the transitional probabilities for Mother Nurse at lag 2, for Mother Groom at lag 3, and for Infant Explore at lags 4 and 5 were significant. Such a pattern of results could suggest the four-event sequence: Infant Active, Mother Touch, Mother Nurse, Mother

Table 7.6. *Results required to confirm the Active, Touch, Nurse, Groom sequence*

Criterion	Target	Lag				
		1	2	3	4	5
Infant Active	Mother Touch	p*	p	p	p	p
	Mother Nurse	p	p*	p	p	p
	Mother Groom	p	p	p*	p	p
Mother Touch	Mother Nurse	p*	p	p	p	p
	Mother Groom	p	p*	p	p	p
Mother Nurse	Mother Groom	p*	p	p	p	p

Note: Asterisks indicate transitional probabilities whose values significantly exceed expected. Numerical values for transitional probabilities have not been given for this hypothetical example.

Groom (we shall return to Infant Explore in a moment), even though the lagged transitional probabilities examined only two codes at a time.

As stated before, the lag sequential is a probabilistic, not an absolute approach. To confirm the putative Active, Touch, Nurse, Groom sequence, we should do the following. First, compute lagged transitional probabilities with Mother Touch as the criterion and Mother Nurse and Mother Groom as targets, then with Mother Nurse as the criterion and Mother Groom as the target. If the transitional probabilities for Mother Nurse at lag 1 and Mother Groom at lag 2, with Mother Touch as the lag 0 criterion, and for Mother Groom at lag 1 with Mother Nurse as the lag 0 criterion, are all significant, then we would certainly be justified in claiming that the Active, Touch, Nurse, Groom sequence was especially characteristic of the monkeys observed (see Table 7.6).

Recall, however, that Infant Explore was significant at lags 4 and 5 after Infant Active. Does this mean that we are dealing with a six-event instead of a four-event sequence? The answer is, not necessarily. For example, if the transitional probabilities for Infant Explore at lags 3 and 4 given Mother Touch as the criterion, at lags 2 and 3 given Mother Nurse, and at lags 1 and 2 given Mother Groom were not significant, then there would be no reasons to claim that Infant Explore followed the Active, Touch, Nurse, Groom sequence already identified. Instead, if the results were as suggested here, we would conclude that after a time when the infant was active, next we would likely see either the Touch, Nurse, Groom sequence or else three more or less random events followed by Infant Explore in the fourth or fifth position.

If we let X stand for a "random" event, then in effect we have detected the following sequence: Active, X, X, X, Explore, Explore. More accurately, because in this example adjacent codes must be different, we have detected the following two sequences: Active, X, X, X, Explore, X and Active, X, X, X, X, Explore. Given other data, we might have detected a sequence like Active, X, Nurse, Groom, which we would interpret as follows: Whatever happens after times when the infant is active is not systematic – it could be almost any code, randomly chosen. After a random event in lag 1, however, the Nurse, Groom sequence is likely (in lag positions 2 and 3). Such a sequence would not be easily detected with "absolute" methods. One advantage, then, of the lag sequential approach is the ease with which sequences containing random elements can still be detected. The main advantage of this approach, however, remains its ability to detect sequences involving more than two events without requiring as much data as absolute methods would. When interpreting lag sequential results, a number of cautions apply. First, it is important to keep in mind whether adjacent codes can be the same or not, because this affects how expected frequencies or probabilities are computed. In the previous section, we noted that when consecutive codes cannot repeat, expected frequencies for lag 1 are best computed using an iterative procedure, although Equation 7.7 from the lag sequential literature provides an approximation. A similar approximation is suggested by Sackett (1979) for lags greater than 1. It is

$$ m_{GT} = \frac{x_T - x_{GT}}{N_1 - x_G} x_G \tag{7.8} $$

where m_{GT} represents the expected frequency for the target behavior at lag L when preceded by the given behavior at lag 0, and x_{GT} the observed frequency for the target at lag $L - 1$ preceded by the given behavior at lag 0.

Sackett (1979) reasoned that when adjacent codes cannot repeat, the expected probability for a particular target code at lag L (assuming a particular given code at lag 0) is the frequency for that target code diminished by the number of times it appears in the lag $L - 1$ position (because then it could not appear in the L position, after itself) divided by the number of events that may occur at lag L (which is the sum of the lag L minus the lag $L - 1$ frequencies summed across all K target codes). Simply put, this sum is the number of all events less the number of events assigned the given code. As with Equation 7.7, Equation 7.8 assumes overlapped sampling; and again like Equation 7.7, marginals for expected frequencies based on Equation 7.8 do not match the observed marginals.

Nonetheless, when consecutive codes cannot repeat, traditional lag sequential analysis (Sackett, 1979; the first edition of this book) has estimated expected frequencies at lag 1 with Equation 7.7 and at longer lags with Equation 7.8, and then has determined statistical significance based on the z computed per Allison and Likers's (1982) Equation 7.6. As already noted, at lag 1 we recommend the log-linear methods described in the previous section, and in the next section we develop log-linear methods that apply at longer lags. But note, when consecutive codes may repeat, and when overlapped sampling is used, at lag 1 traditional lag sequential (Equations 7.4 and 7.6) and log-liner (Equations 7.1 and 7.2) analyses produce almost identical results. The same is essentially true at longer lags, although then log-linear analyses offers certain additional advantages, as described in the next section.

Two additional cautions should be mentioned. As is always true, no matter the statistic, before significance is assigned to any z score, the investigator should determine that there are sufficient data to justify this (see section 8.5). Finally, as always, the investigator should keep in mind the type I error problem (see section 8.6).

This is probably not a serious problem if sequences beginning with just one, or at most a few, criterion codes are investigated in the context of a confirmatory study. However, if many codes are defined, and if all serve exhaustively as criteria and targets, with many lags, then interpretation of such exploratory results should be guided by the almost certain knowledge that some chance findings are contained therein.

We end this section with a second example of the lag sequential method. For a study of marital communication Gottman, Markman, and Notarius (1977) coded the sequential utterances of a number of nondistressed and distressed couples observed discussing marital problems. Among other questions, these investigators wanted to know what happened after the husband complained about a marital problem. In other words, given Husband Complaint as the criterion code, they computed lagged transitional probabilities for a number of target codes, including Wife Agreement, Wife Complaint, Husband Agreement, and Husband Complaint. (The same code may serve as both criterion and target; however, when adjacent codes must be different, both its observed and expected frequencies at lag 1 will be 0. This is an example of a "structural" zero.) Although 24 codes were defined, the four just listed always included the highest z scores.

An interesting difference was noted between nondistressed and distressed couples. For nondistressed couples, significant z scores occurred only when Wife Agreement (at lags 1, 3, and 5) and Husband Complaint (at lags 2 and 4) served as targets. The process of cycling between Husband Complaint and Wife Agreement, Gottman et al. called "validation." For distressed couples, on the other hand, significant z scores occurred only when Wife Complaint (at lags 1 and 3) and Husband Complaint (at lags 2,

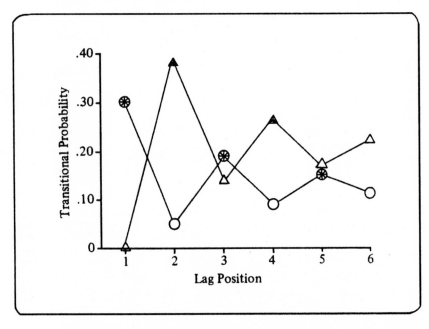

Figure 7.2. A lagged probability profile for nondistressed couples. Triangles represent transitional probabilities for Husband Complaint at the lag specified, given Husband Complaint at lag 0. Circles represent transitional probabilities for Wife Agreement at the lag specified, given Husband Complaint at lag 0. Asterisks (*) indicate that the corresponding z score is significant.

4, and 6) served as targets, a process that Gottman et al. termed "cross-complaining." Lagged probability profiles for these results are presented in Figures 7.2 and 7.3.

7.6 Log-linear approaches to lag-sequential analysis

Since lag-sequential analysis was first developed, log-linear analyses have become more widely understood and used by social scientists (Bakeman & Robinson, 1994; Wickens, 1989). They offer a number of advantages over traditional lag-sequential methods. As already noted, structural zeros are handled routinely and do not require the ad hoc formulas described in the previous section. But primarily, use of log-linear methods allows integration of sequential analysis into an established and well-supported sta-

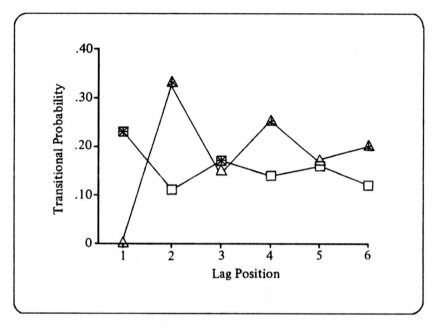

Figure 7.3. A lagged probability profile for distressed couples. As in Figure 7.2, triangles represent transitional probabilities for Husband Complaint at the lag specified, given Husband Complaint at lag 0. Squares, however, represent transitional probabilities for Wife Complaint, given Husband Complaint at lag 0. Again asterisks (*) indicate that the corresponding z score is significant.

tistical tradition (Bakeman & Quera, 1995b). As Castellan (1979) pointed out some time ago, almost always sequential questions can be phrased in terms of multidimensional contingency tables, which log-linear analysis was developed to analyze.

In this section, we describe three advantages of a log-linear view to sequential questions. First, log-linear analysis promotes a whole-table view, whereas often traditional lag-sequential analysis focused, almost piecemeal, on individual transitions in a table, which invites type I error. Moreover, log-linear analysis provides ways of disentangling the web of connected results in a table, as we demonstrate shortly. Finally, log-linear analysis, using well-established techniques, provides an integrated method for determining whether there are effects at various lags, no matter whether consecutive codes may or cannot repeat and no matter whether or not overlapped sampling was employed.

Omnibus tests

Used in an exploratory way, traditional lag-sequential analysis invites type I error (although among statistical techniques it is hardly unique in this respect). When 10 codes are defined, for example, the lag 0 by lag 1 table contains 100 cells when consecutive codes may repeat and 90 when not (K^2 and $K[K - 1]$ generally, where K is the number of codes defined). Assuming the usual .05 value for alpha, if there were no association between lag 0 and lag 1 behavior, approximately 5 of 100 transitions would be identified, on average and incorrectly, as statistically significant. One solution is to take an omnibus or whole-table view. Absent specific predictions that one or just a few transitions will be significant, individual cell statistics should be examined for significance only when a tablewise statistic, such as the Pearson or likelihood-ratio chi-square (symbolized as X^2 and G^2, respectively), is large, just as post hoc tests are pursued in analysis of variance only when the omnibus F ratio is significant (Bakeman, 1992; Bakeman & Quera, 1995b).

Applying this test, we would not have examined the data presented in Table 7.1 further. For these data,

$$X^2(11, N = 127) = \sum_{G}^{K} \sum_{T}^{K} \frac{(x_{GT} - m_{GT})^2}{m_{GT}} = 11.0 \qquad (7.9)$$

and

$$G^2(11, N = 127) = 2 \sum_{G}^{K} \sum_{T}^{K} x_{GT} \log \frac{x_{GT}}{m_{GT}} = 10.3 \qquad (7.10)$$

where *log* represents the natural logarithm (i.e., the logarithm to the base *e*); X^2 and G^2 both estimate chi-square, although usually G^2 is used in log-linear analyses (for a discussion of the differences between them, see Wickens, 1989). These estimates fall short of the .05 critical value of 19.7. Moreover, only 1 of 20 adjusted residuals exceeded 1.96 absolute (Unoccupied to Solitary; see Table 7.5), which suggests it was simply a chance finding, unlikely to replicate.

In log-linear terms, we ask whether expected frequencies generated by the model of independence (i.e., Equation 7.1), which is symbolized [0][1] and indicates the independence of the lag 0 and lag 1 dimension, are similar to the observed frequencies. If they are, then the chi-square statistic will not exceed its .05 critical value, as here (i.e., observed frequencies fit those expected tolerably well). However, if the computed chi-square statistic is large, exceeding its .05 critical value, then we reject the model of independence and conclude that the dimensions of the table are in fact related and not independent.

Table 7.7. *Observed frequencies and adjusted residuals for 100 two-event sequences*

Given code, lag 0	Observed frequencies: Target code, lag 1				Adjusted residuals: Target code, lag 1		
	A	B	C	Totals	A	B	C
A	23	5	15	43	2.02	−1.82	−.056
B	11	1	7	19	1.56	−1.78	−0.12
C	8	14	16	38	−3.32	3.30	0.66
Totals	42	20	38	100			

Note: This example was also used in Bakeman and Quera (1995b).

Winnowing results

As a second example, consider the data given in Table 7.7 for which K is 3; the codes are labeled A, B, and C; and consecutive codes may repeat. For these data, $X^2(4, N = 100)$ is 15.7 and $G^2(4, N = 100)$ is 16.4. Both exceed the .05 critical value of 9.49, which suggests that lag 0 and lag 1 are associated. Moreover, three of nine adjusted residuals exceed 1.96 absolute (again, see Table 7.7). But now we confront a different dilemma. Adjusted residuals in a table form an interrelated web. If some are large, others necessarily must be small, and so, rather than attempting to interpret each one (thereby courting type I error), now we need to determine which one or ones should be emphasized.

The initial set of all statistically significant transitions in a table can be winnowed using methods for incomplete tables (i.e., tables with structural zeros). Assume the C–B chain, whose adjusted residual is 3.30, is of primary theoretical interest. In order to test its importance, we declare the C–B cell structurally zero, use an iterative procedure to compute expected frequencies (e.g., using Bakeman & Robinson's, 1994, ILOG program), and note that now the [0][1] model fits the remaining data ($G^2[3, N = 86] = 5.79$; .05 critical value for 3 $df = 7.81$; $df = 3$ because one is lost to the structurally zero cell). We conclude that interpretation should emphasize the C–B chain as the other two effects (decreased occurrences for C–A, increased occurrences for A–A) disappear when the C–B chain is removed.

Had the model of independence not fit the reduced data, we would have declared another structural zero and tested the data now reduced by two cells. Proceeding stepwise (but letting theoretical considerations not raw empiricism determine the next chain to delete, else one risks capitalizing on chance as with backward elimination in multiple regression and compromising type I error control), we would identify those chains that prevent the [0][1] model from fitting. (A logically similar suggestion, not in a log-linear context, is made by Rechten & Fernald, 1978; see also Wickens, 1989, pp. 251–253.) The ability to winnow results in this way is one advantage of the log-linear view over traditional lag-sequential analysis.

Sequential log-linear models

Perhaps the major advantages of a log-linear approach to sequential problems are the statistical foundation and generality provided. Assuming an interest in lags no greater than L, we begin by assembling $(L + 1)$-event chains. For example, if we were interested in lags no greater than 2, we would collect three-event chains in which the first event was associated with lag 0, the second with lag 1, and the third with lag 2.

The three-event chains might be derived from just one or a few longer sequences using overlapped sampling, selecting first $e_1e_2e_3$, then $e_2e_3e_4$, then $e_3e_4e_5$, etc., where e_i represents an event in the longer sequence. In such cases, if S segments were coded comprising N events in all, the number of three-event chains derived would be $N - SL$ (assuming none of the segments consisted of fewer than $L + 1$ events). For example, in the simplest instance, if one segment consisting of N events were coded, $N - 2$ three-event chains would be tallied. Alternatively, the three-event chains might be derived from a few longer sequences using nonoverlapped sampling (selecting first $e_1e_2e_3$, then $e_4e_5e_6$, etc.), in which case the number of three-event chains derived would be $N \div 3$ (assuming all segments are multiples of three). Or the three-event sequences might be sampled directly from a population of three-event sequences.

Overlapped sampling is often used, and usually assumed, in traditional lag-sequential analysis. But nonoverlapped sampling is both quite common and useful. For example, imagine that we are only interested in tallying speaker 1 (e.g., husband) to speaker 2 (e.g., wife) lag 1 transitions. Then, assuming two-event chains were derived from a single segment of N coded events, the number of speaker 1 to 2 transitions would be $N \div 2$ if the speakers always alternated turns (tallying first e_1e_2, then e_3e_4, etc.), or some smaller number if a speaker's turn may contain more than one thought unit (which could result in tallying e_2e_3, e_5e_6, e_7e_8, $e_{11}e_{12}$, etc.; see Bakeman & Casey, 1995).

No matter the sampling strategy, the $(L + 1)$-event chains are tallied in a K^{L+1} table; thus each chain adds a tally to one, and only one, cell. To demonstrate the log-linear approach, let us begin with an example representing the simplest of circumstances, assuming codes of A, B, and C

1:	A	B	C
0:A	21	25	49
B	23	26	21
C	50	19	15

Figure 7.4. Observed frequencies for 249 two-event chains derived from a sequence of 250 events.

(thus $K = 3$) that may repeat and an initial interest in lag 1. Then occurrences of each of the nine possible two-event chains (AA, AB, etc.) would be tallied in one of the cells of a 3^2 table. For example, we (Bakeman & Quera, 1995b) generated a sequence of 250 coded events and tallied the 249 overlapped two-event chains; the results are shown in Figure 7.4. For this two-dimensional table, the [0][1] model (implying independence of rows and columns) fails to fit the data ($G^2[4, N=249]=35.2$) and so we would conclude that events at lag 0 and lag 1 are associated and not independent. This much seems easy and, apart from the preliminary omnibus test, not much different from traditional lag-sequential methods.

Next, assume that our interest expands from lag 1 to lag 2. Still assuming $K = 3$ and consecutive codes that may repeat, then each of the 27 possible 3-event chains (AAA, AAB, etc.) would be tallied in one of the cells of a 3^3 table. Tallies for the 248 overlapped three-event chains derived from the same sequence used earlier are shown in Figure 7.5.

Cells for the 3^3 table shown in Figure 7.5 are symbolized x_{ijk}, where i, j, and k represent the lag 0, lag 1, and lag 2 dimensions, respectively. Traditional lag-sequential analysis would test for lag 2 effects in the collapsed 02 table, that is, the table whose elements are x_{i+k}, where

$$x_{i+k} = \sum_{j}^{K} x_{ijk}$$

This table is shown in Figure 7.6. For this two-dimensional table, the [0][2] model (implying independence of rows and columns) fails to fit the data ($G^2[4, N = 248] = 10.70$) and so, traditionally, we would conclude that events at lag 0 and lag 2 are associated and not independent. But this fails to take into account events at lag 1.

A hierarchic log-linear analysis of the 3^3 table shown in Figure 7.5 provides more information, and in this case leads to a different conclusion, than a traditional lag-sequential analysis of the collapsed table shown in Figure 7.6. The complete or *saturated* model for a three-dimensional table

		2: A	B	C
0:A	1:A	7	6	8
	B	10	9	6
	C	31	8	10
B	1:A	3	6	14
	B	6	11	9
	C	12	4	4
C	1:A	11	13	26
	B	7	6	6
	C	7	7	1

Figure 7.5. Observed frequencies for 248 three-event chains derived from the same sequence of 250 events used for Figure 7.4.

	2: A	B	C
0:A	48	23	24
B	21	21	27
C	25	26	33

Figure 7.6. Observed frequencies for the collapsed lag 0 × lag 2 table derived from the observed frequencies for three-event chains shown in Figure 7.5.

is represented as [012] and includes seven terms: *012, 01, 12, 02, 0, 1*, and *2*. The saturated model is not symbolized as [012][01][12][02][0][1][2] because the three two-way and three one-way terms are implied by (we could say, nested hierarchically within) the three-way term, and so it is neither necessary nor conventional to write them explicitly.

Typically, a hierarchic log-linear analysis proceeds by deleting terms, seeking the simplest model that nonetheless fits the data tolerably well (Bakeman & Robinson, 1994). Results for the data shown in Figure 7.4 are given in Table 7.8. The best-fitting model is [01][12]; the term that represents lag 0–lag 2 association (i.e., the *02* term) is not required. Thus

Table 7.8. *Hierarchic log-linear analysis of the data shown in Figure 7.4*

Model	G^2	G^2 df	Term Deleted	ΔG^2	ΔG^2 df
[012]	0.0	0			
[01][12][02]	7.67	8	012	7.67	8
[01][12]	11.52	12	02	3.85	4
[01][2]	45.88*	16	12	34.35*	4
[0][1][2]	81.34*	20	01	35.46*	4

**$p < .01$

the log-linear analysis reveals that, when events at lag 1 are taken into account, events at lag 0 and lag 2 are not associated, as suggested by the analysis of the 02 table, but are in fact independent. Such conditional independence – that is, the independence of lag 0 and lag 2 conditional on lag 1 – is symbolized $0 \perp\!\!\!\perp 2 | 1$ by Wickens (1989; see also Bakeman & Quera, 1995b), and the ability to detect such circumstances represents an advantage of log-linear over traditional lag-sequential methods. Readers who wish to pursue the matter of conditional independence further should read Wickens (1989, especially chapter 3).

As just described, when consecutive codes may repeat log-linear but not traditional lag-sequential methods detect conditional independence. Additional advantages accrue when consecutive codes cannot repeat because log-linear methods handle structural zeros routinely and do not require ad hoc and problematic formulas such as Equations 7.7 and 7.8. As an example, we (Bakeman & Quera, 1995b) generated a sequence of 122 coded events and tallied the 120 overlapped three-event chains. Tallies for the 12 permitted sequences are given in Figure 7.7. Cells containing structural zeros are also indicated; when consecutive codes cannot repeat, the 012 table will always contain the pattern of structural zeros shown.

A summary of the log-liner analysis for the data given in Figure 7.7 is shown in Table 7.9. When K is 3, and only when K is 3, the [01][12][02] model is completely determined; its degrees of freedom are 0 and expected frequencies duplicate the observed ones (as in the [012] model, when consecutive codes may repeat). Unlike in the previous analysis, for these data the model of conditional independence – [01][12] – fails to fit the data ($G^2[3, N = 120] = 10.83, p < .05$). Thus we accept the [01] [12] [02] model and conclude that events at lag 0 and lag 2 are associated (and both are associated with lag 1).

	2: A	B	C
0:A 1:A	–	–	–
B	15	–	6
C	10	9	–
B 1:A	–	12	8
B	–	–	–
C	9	12	–
C 1:A	–	8	11
B	5	–	15
C	–	–	–

Figure 7.7. Observed frequencies for the 12 possible three-event chains derived from a sequence of 122 events for which consecutive codes cannot repeat. Structural zeros are indicated with a dash.

Table 7.9. *Hierarchic log-linear analysis of the data shown in Figure 7.7*

Model	G^2	G^2 df	Term Deleted	ΔG^2	ΔG^2 df
[01][12][02]	0.0	0			
[01][12]	10.83*	3	02	10.83*	3
[01][12]–CBC	1.64	2	cell x_{CBC}	9.19*	1

*$p < .05$
**$p < .01$

Moreover, we can winnow these results, exactly as described for the earlier example that permitted consecutive codes to repeat. For theoretic reasons, assume that the CBC chain is of particular interest. An examination of the residuals for the [01] [12] model (i.e., the differences between observed frequencies and expected frequencies generated by the [01] [12] model) showed that 4 of the 12 chains were associated with quite large absolute residuals (the *ABA, ABC, CBA,* and *CBC* chains), which made us think that the observed frequencies for these chains, in particular, might be responsible for the failure of the [01] [12] model to fit the observed data. Because the *CBC* chain is of primary interest, we replaced the x_{CBC} cell (which contained a tally of 15) with a structural zero. As shown in Table 7.9, the model of conditional independence now fit the data ($G^2[2, N = 105] = 1.64$, *NS*), and so we conclude that the *CBC* chain can account for the lag 2 effect detected by the omnibus analysis described in the previous paragraph. This can be tested directly with a hierarchic test, as indicated in Table 7.9. The difference between two hierarchically related G^2s is distributed approximately as chi-square with degrees of freedom equal to the difference between the degrees of freedom for the two G^2s; in this case, $\Delta G^2(1) = 9.19$, $p < .01$. (Replacing a different chain with a structural zero might also result in a fitting model, which is why it is so important that selection of the chain to consider first be guided by theoretic concerns.)

Minimizing data demands

Quantitative data analysis always requires sufficient data, and log-linear and traditional lag-sequential approaches are no exception. Still, the data required for the multidimensional tables of log-linear analysis can be quite intimidating. Several rules of thumb for log-linear analysis are available, usually stated in terms of expected frequencies or even degrees of freedom for hierarchic tests like the one for cell x_{CBC} shown in Table 7.9, but one suggested requirement (a necessary minimum, but not necessarily sufficient) is that the total sample be at least 4 or 5 times the number of cells not structurally zero (Wickens, 1989, p. 30). This number is K^{L+1} when consecutive codes may repeat and $K(K-1)^L$ when they cannot (e.g., when $K = 3$ and $L = 2$, the number of cells is 27 and 12 when codes may and cannot repeat, respectively). Thus the number of cells, and so the total sample desired, increases exponentially with increases in K and L (although the increase is somewhat less pronounced when consecutive codes cannot repeat).

Especially for larger values of L, unless the number of events observed is almost astronomically large, the average number of events per cell may be distressingly small. Further, expected frequencies for far too many of the cells may be near zero, which is problematic for log-linear analysis. To minimize data demands, Bakeman and Quera (1995b) have suggested a sequential search strategy for explicating lagged effects. Although the details are somewhat different, the general strategy is the same when consecutive codes may and cannot repeat, which once again demonstrates the generality of the log-linear approach.

Consider first the strategy when consecutive codes may repeat. Lag 1 effects, which require only the two-dimensional 01 table, are unproblematic, although of course we would test for lag 1 effects. Thus the search begins by looking for complex lag 2 effects in the 012 table. At each lag ($L > 1$), the complex effects we seek first implicate lags 0 and L with lag $L - 1$. If present, collapsing over the $L - 1$ dimension (which would reduce the number of cell and so data demands) is unwarranted. For example, if L is 2, then complex effects are present if the simplest model that fits the 012 table includes any of the following:

1. [012] because then lag 0 and lag 2 are associated and interact with lag 1 (three-way associations), or
2. [01][12][02] because then lag 0 and lag 2 are associated with each other and lag 1 but do not interact with lag 1 (homogeneous associations), or
3. [01][12] because then lag 0 and lag 2 are independent conditional on lag 1.

If complex effects are found, we would explicate them, as demonstrated in the previous section. However, if simpler models fit (e.g., [01][2] or any others not in the list just given), which means no complex effects were found, then collapsing over the $L - 1$ dimension is justified (Wickens, 1989, pp 79–81, pp. 142–143), resulting in the 0L table of traditional lag-sequential analysis (e.g., the 02 table when $L = 2$).

Assuming no complex effects are found in the 012 table, after collapsing we would first test whether $0 \perp\!\!\!\perp 2$ (unconditional independence) in the 02 table and, if not, examine residuals in order to explicate the lag 0–lag 2 effect just identified (exactly as we would have done for the 01 table). Next we would create a new three-dimensional table by adding the lag 3 dimension, tally sequences in this 023 table, and then look for lag 3 effects in the 023 table exactly as described for the 012 table. This procedure is repeated for successive lags. In general terms, beginning with lag L, we test whether the three-dimensional $0(L - 1)L$ table can be collapsed over the $L - 1$ dimension. If so, we collapse to the 0L table, add the $L + 1$ dimension thereby creating a new three-dimensional table, increment L, and repeat the procedure, continuing until we find a table that does not permit collapsing over the $L - 1$ dimension. Once such a table is found, we explicate the lag L effects in this three-dimensional table. If data are sufficient, we might next analyze the four-dimensional $0(L - 1)L(L + 1)$ table, and so forth, but further collapsing is unwarranted because of the lag L effects just found. Nonetheless, this strategy may let us examine lags longer than 2 without requiring tables larger then K^3 when consecutive codes may repeat.

The sequential search strategy described in the previous paragraph applies when consecutive codes cannot repeat with one modification. When consecutive codes may repeat, and no complex lag L effects are found (i.e., each table examined sequentially permits collapsing) then the test series becomes 0⊥2, then 0⊥3 and so forth (i.e., 0⊥L is tested in the 0L table), as just described. When consecutive codes cannot repeat, the unconditional test makes no sense because it fails to reflect the constraints imposed when consecutive codes cannot repeat. Then, when no complex lag L effects are found, the analogous series becomes 0⊥2|1, 0⊥3|2, and so forth [i.e., 0⊥L|$L - 1$ is tested in the 0($L - 1$)L table]. Models associated with these tests include the $(L - 1)L$ term. The corresponding marginal table has structural zeros on the diagonal, which reflect the cannot-repeat constraint. This strategy may let us examine lags longer than 2 without requiring tables larger than $K^2(K - 1)$ when consecutive codes cannot repeat $(K[K - 1]^2$ when $L = 2)$. These matters are discussed further in Bakeman and Quera (1995b).

7.7 Computing Yule's Q or phi and testing for individual differences

Often more than a single individual, dyad, family, or whatever, is observed; these *units* are embedded in a design (e.g., a two-group design might include clinic and nonclinic couples), and investigators want to ask questions about the importance of their research factors (e.g., is a particular sequential pattern more characteristic of clinic than nonclinic couples). The previous edition of this book suggested than z scores might serve as scores for subsequent analyses (e.g., analyses of variance, multiple regression, etc.), but that was not sound advice. The z score is affected by the number of tallies (if the number of tallies doubled but the association remained the same, the z score would increase), and so is not comparable across experimental units (subjects, dyads, families, etc.) unless the total number of tallies remains the same for each. Some measure that is unaffected by the number of tallies, such as a strength of association or effect size measure, should be used instead (Wampold, 1992).

Strength of association or effect size measures are especially well developed for 2 × 2 tables (to give just two examples from an extensive literature, see Conger & Ward, 1984, and Reynolds, 1984; much of the material in this and subsequent paragraphs is summarized from Bakeman, McArthur, & Quera, 1996). This is fortunate, because when interest centers on one cell in a larger two-dimensional table, the larger table can be collapsed into a 2 × 2, and statistics developed for 2 × 2, tables can be used (as Morley, 1987, noted with respect to phi). Assume, for example, that we want to know whether event B is particularly likely after event A. In this case, we would label rows A and $\sim A$ and columns B and $\sim B$ (where rows represent lag 0, columns lag 1, and $\sim$ represents *not*). Then the collapsed 2 × 2 table can be represented as

	B	~ B
A	a	b
~ A	c	d

where individual cells are labeled a, b, c, and d as shown and represent cell frequencies.

One of the most common statistics for 2×2 tables (perhaps more so in epidemiology and sociology than psychology) is the odds ratio. As its name implies, it is estimated by the ratio of a to b divided by the ratio of c to d,

$$est. \; odds \; ratio = \frac{a/b}{c/d} \tag{7.11}$$

(where a, b, c, and d refer to observed frequencies for the cells of a 2×2 table as noted earlier; notation varies, but for definitions in terms of population parameters, see Bishop Fienberg, & Holland, 1975; and Wickens, 1993). Multiplying numerator and divisor by d/c, this can also be expressed as

$$est. \; odds \; ratio = \frac{ad}{bc}. \tag{7.12}$$

Equation 7.12 is more common, although Equation 7.11 reflects the name and renders the concept more faithfully. Consider the following example:

	B	~ B	
A	10	10	20
~ A	20	60	80
	30	70	100

The odds for B after A are 1:1, where as the odds for B after any other (non-A) event are 1:3; thus the odds ratio is 3. In other words, the odds for B occurring after A are three times the odds for B occurring after anything else. When the odds ratio is greater than 1 (and it can always be made ≥ 1 by swapping rows), it has the merit, lacking in many indices, of a simple and concrete interpretation.

The odds ratio varies from 0 to infinity and equals 1 when the odds are the same for both rows (indicating no effect of the row classification). The natural logarithm (ln) of the odds ratio, which is estimated as

$$est. \; log \; odds \; ratio = \ln \left(\frac{ad}{bc} \right) \tag{7.13}$$

extends from minus to plus infinity, equals 0 when there is no effect, and is more useful for inference (Wickens, 1993). However Equation 7.13 estimates are biased. An estimate with less bias, which is also well defined when one of the cells is zero (recall that the log of zero is undefined), is obtained by adding 1/2 to each count,

$$y = \ln \frac{(a + 1/2)(d + 1/2)}{(c + 1/2)(b + 1/2)} \tag{7.14}$$

(Gart & Zweifel, 1967; cited in Wickens, 1993, Equation 8). As Wickens (1993) notes when recommending that the log odds ratio computed per Equation 7.14 be analyzed with a parametric *t* test, this procedure not only provides protection for a variety of hypotheses against the effects of intersubject variability when categorical observations are collected from each member of a group (or groups), it is also easy to describe, calculate, and present.

Yule's Q

Yule's Q is a related index. It is a transformation of the odds ratio designed to vary, not from zero to infinity with 1 indicating no effect, but from -1 to $+1$ with zero indicating no effect, just like the familiar Pearson product – moment correlation. For that reason many investigators find it more descriptively useful than the odds ratio. First, c/d is subtracted from the numerator so that Yule's Q is zero when a/b equals c/d. Then, a/b is added to the denominator so that Yule's Q is $+1$ when b and/or c is zero and -1 when a and/or d is zero, as follows:

$$\text{Yule's Q} = \frac{\dfrac{a}{b} - \dfrac{c}{d}}{\dfrac{c}{d} + \dfrac{a}{b}} = \frac{\dfrac{ad - bc}{bd}}{\dfrac{bc + ad}{bd}} = \frac{ab - bc}{ad + bc} \tag{7.15}$$

Yule's Q can be expressed as a monotonically increasing function of both the odds and log odds ratio; thus these three indices are equivalent in the sense of rank ordering subjects the same way (Bakeman, McArthur, & Quera, 1996).

Phi

Another extremely common index for 2 × 2 tables is the phi coefficient. This is simply the familiar Pearson product–moment correlation coefficient computed using binary coded data (Cohen & Cohen, 1983; Hays, 1963).

One definition for phi is

$$\phi = \frac{z}{\sqrt{N}} \qquad (7.16)$$

where z is computed for the 2×2 table and hence equals $\sqrt{\chi^2}$. Thus phi can be viewed as a z score corrected for sample size. Like Yule's Q, it varies from -1 to $+1$ with zero indicating no association. In terms of the four cells, phi is defined as

$$\phi = \frac{ad - bc}{\sqrt{(a + b)(c + d)(a + c)(b + d)}} \qquad (7.17)$$

Multiplying and rearranging terms this becomes

$$\phi = \frac{ad - bc}{\sqrt{(ac + bd + ad + bc)(ab + cd + ad + bc)}} \qquad (7.18)$$

If we now rewrite the expression of Yule's Q, first squaring the denominator of Equation 7.15 and then taking its square root

$$\text{Yule's Q} = \frac{ad - bc}{\sqrt{(ad + bc)(ad + bc)}} \qquad (7.19)$$

the value of Yule's Q is not changed but similarities and differences between phi and Yule's Q (Equations 7.18 and 7.19) are clarified.

Does it matter which index is used, Yule's Q or phi? The multiplier and multiplicand in the denominator for Yule's Q (Equation 7.19) consist only of the sum of ad and bc, whereas multiplier and multiplicand in the phi denominator (Equation 7.18) add more terms. Consequently, values for phi are always less than values for Yule's Q (unless b and c, or a and d, are both zero, in which case both Yule's Q and phi would be $+1$ and -1, respectively). Yule's Q and phi differ in another way as well. Yule's Q is $+1$ when either b or c is zero and -1 when either a or d is zero (this is called weak perfect association, Reynolds, 1984), whereas phi is $+1$ only when both b and c are zero and -1 only when both a and d are zero (this is called strict perfect association). Thus phi achieves its maximum value (absolute) only when row and column marginals are equal (Reynolds, 1984). Some investigators may regard this as advantageous, some as disadvantageous, but in most cases it probably matters little which of these two indices is used (or whether the odds ratio or log odds ratio is used instead). In fact, after running a number of computer simulations, Bakeman, McArthur, and Quera (1996) concluded that, when testing for group differences, it does not matter much whether Yule's Q or phi is used since both rank-order cases essentially the same. Transformed kappa, a statistic proposed by Wampold

(1989, 1992), however, did not perform as well. For details see Bakeman, McArthur, and Quera (1996).

Type I error considerations

In an ideal confirmatory world, investigators would pluck the one transition from a larger table needed to answer their most important research question; compute a single Yule's Q or phi based on a collapsed A, $\sim$ A|B, $\sim$ B table, such as the one shown earlier; and proceed to test for group differences (or other questions as their design permits). But much of the world is rankly exploratory. Indeed, it is tempting to compute some index for each of the K^2 cells of a table ($K[K-1]$ cells when consecutive codes cannot repeat), one set for each subject, and then subject all K^2 scores to standard parametric tests (t test, analyses of variance, etc.). This courts type I error in a fairly major way. At the very least, no more indices should be derived than the degrees of freedom associated with the table, which is $(K-1)(K-1)$ when consecutive codes may repeat and $(K-1)(K-1)-K$ when not (assuming a table with structural zeros on the diagonal). This is somewhat analogous to decomposing an omnibus analysis of variance into single-degree-of-freedom planned comparisons or contrasts.

One systematic way to derive indices from a larger table requires that one code be regarded as something of a baseline, or base for comparison, such as *unengaged* or *no activity*. For example, imagine that codes are labeled *A, B,* and *C*, and that code *C* represents some sort of baseline. Then following Reynolds's (1984) suggestion for decomposing the odds ratio in tables larger than 2 × 2, and labeling the cells in the 3^2 table as follows:

	A	B	C
A	*a*	*b*	*c*
B	*d*	*e*	*f*
C	*g*	*h*	*i*

four 2 × 2 tables would be formed for each subject, as follows:

	A	C
A	*a*	*c*
C	*g*	*i*

	B	C
A	*b*	*c*
C	*h*	*i*

B	*d*	*f*
C	*g*	*i*

B	*e*	*f*
C	*h*	*i*

and a Yule's Q or phi computed for each. These statistics could then be subjected to whatever subsequent analyses the investigator deems appropriate.

In this section we have suggested that sequential associations between two particular events (e.g., an A to B transition) be assessed with an index like Yule's Q or phi. These statistics gauge the magnitude of the effect and, unlike the z score, are unaffected by the number of tallies. Thus they are reasonable candidates for subsequent analyses such as the familiar parametric tests routinely used by social scientists to assess individual differences and effects of various research factors (e.g., t tests, analyses of variance, and multiple regression). But the events under consideration may be many in number, leading to many tests and thereby courting type I error.

It goes without saying (which may be why it is so necessary to restate) that guiding ideas provide the best protection against type I error. Given K codes and an interest in lag 1 effects, a totally unguided and completely exploratory investigator might examine occurrences of all possible K^2 two-event chains (or $K[K-1]$ two-event chains when consecutive codes cannot repeat). In this section, we have suggested that a more justifiable approach would limit the number of transitions examined to the $(K-1)^2$ degrees of freedom associated with the table (or $[K-1]^2 - K$ degrees of freedom when consecutive codes cannot repeat) and have demonstrated one way that this number of 2×2 subtables could be extracted from a larger table. Presumably a Yule's Q or some other statistic would be computed for each subtable. Positive values would indicate that the pair of events associated with the upper-left-hand cell is associated more than expected, given the occurrences observed for the baseline events associated with the second row and second column of the 2×2 table. The summary statistic for the 2×2 tables, however many are formed, could then be subjected to further analysis.

Investigators are quite free – in fact, encouraged – to investigate a smaller number of associations (i.e., form a smaller number of 2×2 tables). For example, a larger table might be collapsed into a smaller one, combining some codes that seem functionally similar, or only those associations required to address the investigator's hypotheses might be subjected to analysis in the first place. Other transitions might be examined later, and those analyses labeled exploratory instead of confirmatory. For further discussion of this "less is more" and "least is last" strategy for controlling type I error, see Cohen and Cohen (1983, pp. 169–172).

7.8 Summary

Investigators often represent their data as sequences of coded events. Sometimes, data are recorded as event sequences in the first place; other times, in order to answer particular questions, event sequences are extracted from data initially recorded and represented in some other way. The purpose of this chapter has been to describe some ways of analyzing such event sequences, although much of what has been presented here can apply to the analysis of time sequences as well.

Sometimes consecutive events cannot be assigned the same code in event sequences. For example, when coders are asked to segment the stream of behavior into mutually exclusive and exhaustive behavioral states, often adjacent states cannot be coded the same way, by definition. It they were the same, they would be just one state. However, we can imagine other ways of defining event boundaries that would allow adjacent codes to be the same. (The codes used to categorize events would still be mutually exclusive and exhaustive, but that is a different matter.) For example, if utterances were being coded, two successive utterances might both be coded the same. Whether adjacent codes can be the same or not is an important matter because it affects the way expected frequencies, expected probabilities, and hence adjusted residuals (i.e., z scores) are computed.

One approach to sequence detection we have called "absolute." Investigators define particular sequences of some specified length, categorize and tally all sequences of that length, and report the frequencies and probabilities for particular sequences. A z score can be used to gauge the extent to which an observed frequency (or probability) for a particular sequence exceeds its expected value. However, if the z score is to be tested for significance, its computation should be based on sufficient data to justify the normal approximation to the binomial distribution.

In theory, absolute methods apply to sequences of any length. In practice, certain limitations may prevail. In particular, the number of possible sequences increases dramatically as longer and longer sequences are considered. Unless truly mammoth amounts of data are available, expected frequencies for a particular sequence may be too small to justify assigning significance. Moreover, the number of occurrences for a particular sequence may be so few that the investigator has little confidence in the accuracy of the observed frequency, even descriptively. Another exacerbating circumstance is the number of codes defined. In general, when there are more codes, the expected frequencies for particular sequences are likely to be smaller, and hence more data will be required.

Even when z-score computations are based on sufficient data, the type I error problem remains. This is usually not a problem for confirmatory studies, assuming, of course, that just a few theoretically relevant tests of significance are made. But when the number of tests is large, as it typically is for exploratory studies, then some thought should be given to ways to control the type I error rate. As discussed in this chapter, the number of initially significant transitions can be winnowed using structural zeros and log-linear methods, and the number of transitions examined in the first place can be limited to those of clear theoretic interest. Nonetheless, interpretation of results may need to take into account that some of the apparently significant findings are due simply to chance.

One way to describe two-event sequences is to report their simple probabilities. Another way is to report (lag 1) transitional probabilities instead. Of the two, transitional probabilities (which "control" for differences in the base rate of the first or "given" code) often seem more informative descriptively. The values of both, however, are affected by the values for the base rates of the two codes involved. This does not affect their descriptive value, but in cases in which transitional probabilities have been computed separately for different subjects, it does make such scores poor candidates for subsequent analyses of individual or group differences. First, depending on base rates, similar numerical values for the same transitional probability may have quite different meanings for different subjects. And second, what appear to be analyses of transitional probabilities may in fact reflect little more than simple probability effects. The z scores are not hampered by these problems, but have additional problems of their own. Their values are affected by the number of tallies, and so larger values may reflect, not a larger effect, but simply more data. Whenever individual or group differences or effects of research factors generally are of interest, magnitude of effect statistics, and not z scores, should be used. Examples include the odds ratio, the log odds ratio, Yule's Q, and phi.

A second approach to sequence detection, the lag-sequential method, we have characterized as "probabilistic" instead of "absolute." Like that of absolute methods, its purpose is to detect commonly occurring sequences, but because it examines codes pairwise only, it can detect sequences longer than two events without invoking the same restrictions and limitations involved with absolute methods. Moreover, sequences containing random elements can be detected as well. The method is based on an examination of z scores associated with transitional probabilities computed for various lags; thus any limitations and cautions (including the type I error problem) that apply to two-event transitional probabilities also apply to the lag-sequential approach as well.

Perhaps the most adequate approach to sequence detection is log-linear. Log-linear analysis promotes a whole-table view, whereas often traditional lag-sequential analysis focused, almost piecemeal, on individual transitions in a table. This is not necessarily problematic when, in the context of a confirmatory study, only a few transitions are of interest, but a narrow focus on repeated tests tied to all cells in the context of an exploratory study invites type I error. Additionally, log-linear analysis provides ways of disentangling the web of connected results in a table, and makes routine the analysis of sequences in which, for logical reasons, consecutive codes cannot repeat. Finally, log-linear analysis, using well-established statistical techniques, provides an integrated method of broad generality for determining whether there are effects at various lags, no matter whether consecutive codes may or cannot repeat and no matter whether or not overlapped sampling was employed. Whenever possible, it seems the analytic approach of choice for the analysis of coded sequences.

8

Issues in sequential analysis

8.1 Independence

In classical parametric statistics, we assume that our observations are independent, and this assumption forms part of the basis of our distribution statistics. In the sequential analysis of observational data, on the other hand, we want to *detect dependence* in the observations. To do this we compare observed frequencies with those we would expect if the observations were independent. Thus, dependence is not a "problem." It is what we are trying to study.

The statistical problem of an appropriate test is not difficult to solve. It was solved in a classic paper in 1937 by Anderson and Goodman (see also Goodman, 1983, for an update). Their approach is based on the likelihood-ratio chi-square test.

The likelihood-ratio test compares the goodness of fit of any two statistical models if one (the "little" model) is a subset of the other (the "big" model). The null-hypothesis model is often the little model. In our case, this model is often the assumption that the codes are independent (or quasi independent); i.e., that there is no sequential structure. Compared to this is the big, interesting model that posits a dependent sequential structure. As discussed in section 7.6, the difference between the G^2 for the big model (e.g., [01]) and the G^2 for the little model (e.g., [0][1]) is distributed asymptotically as chi-square, with degrees of freedom equal to the difference in the degrees of freedom for the big and little models. "Asymptotic" means that it becomes increasingly true for large N, where N is the number of observations.

When the data have "structural zeros," e.g., if a code cannot follow itself (meaning that the frequency for that sequence is necessarily zero), the number of degrees of freedom must be reduced (by the number of cells that are structural zeros). These cells are not used to compute chi-square (see Goodman, 1983).

We shall now discuss the conditions required to reach asymptote. In particular, we shall discuss assigning probability values to z scores. We

136

should note that most observational data are *stochastically* dependent. They are called in statistics "*m*-dependent" processes, which means that the dependencies are short lived. One implication of this is that there is poor predictability from one time point to another, as the lag between time points increases. In time-series analysis, forecasts are notoriously poor if they exceed one step ahead (see Box & Jenkins, 1970, for a graph of the confidence intervals around forecasts). It also means that clumping *m* observations gives near independence. For most data, *m* will be quite small (probably less than 4), and its size relative to *n* will determine the speed at which the asymptote is approached.

We conclude that assigning probability values to pairwise *z* scores (or tablewise chi-squares) is appropriate when we are asking if the observed frequency for a particular sequence is significantly different from expected (or whether lag 0 and *L* are related and not independent). We admit, however, that more cautious interpretations are possible, and would quote a paragraph we wrote earlier (Gottman & Bakeman, 1979, p. 190):

> As *N* increases beyond 25, the binomial distribution approximates a normal distribution and this approximation is rapidly asymptotic if *P* is close to 1/2 and slowly asymptotic when *P* is near 0 or 1. When *P* is near 0 or 1, Siegel (1956) suggested the rule of thumb that $NP(1 - P)$ must be at least 9 to use the normal approximation. Within these constraints the *z*-statistic above is approximately normally distributed with zero mean and unit variance, and hence we may cautiously conclude that if *z* exceeds ± 1.96 the difference between observed and expected probabilities has reached the .05 level of significance (see also Sackett, 1978). However, because dyadic states in successive time intervals (or simply successive dyadic states in the case of event-sequence data) are likely not independent in the purest sense, it seems most conservative to treat the resulting *z* simply as an index or score and not to assign *p*-values to it.

As the reader will note, in the chapter just quoted we were concerned with two issues: the assumption of independence and the number of tallies required to justify use of the binomial distribution. On reflection, we find the argument that the categorizations of successive *n*-event sequences in event-sequence data are not "independent" less compelling than we did previously, and so we are no longer quite so hesitant to assign probability values on this score.

Our lack of hesitancy rests in part on a simulation study Bakeman and Dorval (1989) performed. No matter the statistic, for the usual sorts of parametric tests, *p* values are only accurate when assumptions are met. To those encountering sequential analysis for the first time, the common (but not necessarily required) practice of overlapped sampling (tallying first the e_1e_2 chain, then e_2e_3, e_3e_4, etc.) may seem like a violation of independence. The two-event chain is constrained to begin with the code that ended the previous chain (i.e., if a two-event chain ends in B, adding a tally to the 2nd column, the next must add a tally to the 2nd row), and this violates sampling independence. A Pearson or likelihood-ratio chi-square could be computed and would be an index of the extent to which observed frequencies in this table tend to deviate from their expected ones. But we would probably have even less confidence than usual that the test statistic is distributed as χ^2 and so, quite properly, would be reluctant to apply a p value.

Nonoverlapped sampling (tallying first the $e_1 e_2$ chain, then $e_3 e_4$, $e_5 e_6$, etc.) does not pose the same threat to sampling independence, although it requires sequences almost twice as long in order to extract the same number of two-event chains produced by overlapped sampling. However, the consequences of overlapped sampling may not be as severe as they at first seem. Bakeman and Dorval (1989) found that when sequences were generated randomly, distributions of a test statistic assumed their theoretically expected form equally for the overlapped and nonoverlapped procedures and concluded that the apparent violation of sampling independence associated with overlapped sampling was not consequential.

8.2 Stationarity

The term "stationarity" means that the sequential structure of the data is the same independent of where in the sequence we begin. This means that, for example, we will get approximately the same antecedent/consequent table for the first half of the data as we get for the second half of the data.

	1st Half		2nd Half	
	HNice	HNasty	HNice	HNasty
WNice	80	10	64	7
WNasty	2	60	12	66

We then compute the pooled estimates over the whole interaction:

	HNice	HNasty
WNice	144	17
WNasty	14	126

To test for stationarity of the data, we compare the actual data to the expected values under the null hypothesis that the data are stationary. Let's assume that we are interested in only the lag-1 antecedent/consequent table for s codes. Let $N(IJ, t)$ be the joint frequency for cell IJ in segment t, and

$P(IJ, t)$ be the transition probability for that cell. Let $P(IJ)$ be the pooled transition probability. Then G^2, computed as

$$G^2 = 2 \sum_t N(IJ, t) \log_e \left(\frac{P(IJ, t)}{P(IJ)} \right)$$

is distributed as chi-square. If there are s codes, and T segments, then G^2 has degrees of freedom $(T - 1)(s)(s - 1)$. A more general formula for the rth-order transition table is given by Gottman & Roy (1990, pp. 62–63), where r is the order of the chain. The sum is across segments of the data. For the example data, the value computed was 6.44, which is compared to $df = 2$; this quantity is not statistically significant. Hence, the example data are stationary.

This test can be used to see if the data have a different sequential structure for different parts of the interaction. This appears to be the case for conflict resolution in married couples; the first third is called the "agenda building segment," the second third is called the "disagreement segment," and the final third is called the "negotiation segment" (Gottman, 1979a). However, a problem with this test is that as the number of observations increases, the power we have to detect violations of absolute stationarity increases, and yet, for all intents and purposes the data may be stationary enough.

In this case we can do a log-linear analysis of the data and evaluate the Q^2 statistic, as recommended by Bakeman and Robinson (1994, p. 102 ff.). For example, let C = consequent (husband nice/husband nasty), A = antecedent (wife nice/wife nasty), and T = segment (first half/second half). If the CAT term is required for a fitting model, this suggests that the association between antecedent and consequent varies as a function of time (i.e., is not stationary). For the present data, the loss in fit when this term is removed is significant at the .10 but not the .05 level ($G^2[1] = 3.11$). The G^2 for the base model (i.e., [C][A][T]) is 227.4, so the 3.11 represents less than 1.4% of the total. Even when the loss in fit is significant at the .05 or a more stringent level, Knoke and Burke (1980) recommend ignoring terms that account for less than 10% of the total. This can be useful when even small effects are statistically significant, as often happens when the number of tallies is large.

8.3 Describing general orderliness

The material already presented in the last chapter assumes that investigators want to know how particular behavioral codes are ordered. For example, they may want to confirm that a particular, theoretically important sequence (like Touch, Nurse, Groom) really occurs more often than expected, given base rates for each of these codes. Or, in a more exploratory vein, they may want to identify whichever sequences, if any, occur at greater than chance rates. However, there is another quite different kind of question investigators can ask, one that concerns not how particular codes in the stream of behavior are ordered, but how orderly the stream of behavior is overall.

In this section, we shall not describe analyses of overall order in any detailed way. Instead, we shall suggest some references for the interested reader, and shall try to give a general sense of what such analyses reveal and how they proceed. Primarily, we want readers to be aware that it is possible to ask questions quite different from those discussed earlier in this chapter.

One traditional approach to the analysis of general order is provided by what is usually called "information theory." A brief explication of this approach, along with appropriate references and examples, is given by Gottman and Bakeman (1979). Although the classical reference is Shannon and Weaver (1949), more useful for psychologists and animal behaviorists are Attneave (1959) and Miller and Frick (1949). A well-known example of information theory applied to the study of social communication among rhesus monkeys is provided by S. Altmann (1965). A closely related approach is called Markovian analysis (e.g., Chatfield, 1973). More recently, problems of gauging general orderliness are increasingly viewed within a log-linear or contingency-table framework (Bakeman & Quera, 1995b; Bakeman & Robinson, 1994; Bishop, Fienberg, & Holand, 1975; Castellan, 1979; Upton, 1978).

No matter the technical details of these particular approaches, their goals are the same: to determine the level of sequential constraint. For example, Miller and Frick (1949), reanalyzing Hamilton's (1916) data concerning trial-and-error behavior in rats and 7-year-old girls, found that rats were affected just by their previous choice whereas girls were affected by their previous two choices. In other words, if we want to predict a rat's current choice, our predictions can be improved by taking the previous choice into account but are not further improved by knowing the choice before the previous one. With girls, however, we do improve predictions concerning their current choice if we know not just the previous choice but the one before that, too.

If data like these had been analyzed with a log-linear (or Markovian) approach, the analysis might have proceeded as follows: First we would define a zero-order or null model, one that assumed that all codes occurred with equal probability and were not in any way sequentially constrained. Most likely, the data generated by this model would fail to fit the observed data. Next we would define a model that assumed the observed probabilities for the codes but no sequential constraints. Again, we would test whether the data generated by this model fit the observed. If this model failed to fit, we would next define a model that assumed that codes are constrained just by the immediately previous code (this is called a first-order Markov process). In terms of the example given above, this model should generate data that fit those observed for rats but not for girls. Presumably, a model that assumes that codes are constrained by the previous two codes should generate data that pass the "fitness test" for the girl's data.

In any case, the logic of this approach should be clear. A series of models are defined. Each imposes an additional constraint, for example, that the data generated by the model need to take into account the previous code, the previous two codes, etc. The process stops when a particular model generates data similar to what was actually observed, as determined by a goodness-of-fit test. The result is knowledge about the level of sequential constraint, or connectedness, or orderliness of the data, considered as a whole. (For a worked example, analyzing mother–infant interaction, see Cohn & Tronick, 1987.)

8.4 Individual versus pooled data

In the previous chapter, we discussed two different uses of sequential statistics such as z scores (i.e., adjusted residuals and Yule's Q's). First, assuming that successive codings of events are independent of previous codings, and assuming that enough data points are available, we have suggested that z scores can be tested for significance (see section 7.4). Second, when data are collected from several different "units" (e.g., different participants, dyads, or families), we have suggested that scores such as Yule's Q can be used in subsequent analyses of individual or group differences (see section 7.7). Because the familiar parametric techniques (e.g., analyses of variance) are both powerful and widely understood, such a course has much to recommend it.

Not all studies include readily discernible "units" however. For example, just one individual or couple might be observed, or the animals observed might not be easily identifiable as individuals. In such cases, the issue of pooling data across units does not arise; there are not multiple units. In other cases, for example, when several different children are observed, so few data might be collected for each child that pooling data across all children could seem desirable, if for no other reason than to increase the reliability of the summary statistics reported. At the same time, assuming enough data, it might then be possible to test z scores for significance on the basis of the pooled data. Properly speaking, however, any conclusions from such analyses should be generalized just to other behavior of the group observed, not to other individuals in the population sampled.

Thus, even though investigators who pool data over several subjects usually do so for practical reasons, it has some implications for how results are interpreted.

How seriously this last limitation is taken seems to vary somewhat by field. In general, psychologists studying humans seem reluctant to pool data over subjects, often worrying that some individuals will contribute more than others, thereby distorting the data. Animal behaviorists, on the other hand, seem to worry considerably less about pooling data, perhaps because they regard their subjects more as exemplars for their species and focus less on individuality. Thus students of animal behavior often seem comfortable generalizing results from pooled data to other members of the species studied.

As we see it, there are three options: First, when observations do no derive from different subjects (using "subject" in the general sense of "case" or "unit"), the investigator is limited to describing frequencies and probabilities for selected sequences. Assuming enough data, these can be tested for significance. Second, even when observations do derive from different subjects, but when there are few data per subject, the investigator may opt to pool data across subjects. As in the first case, sequences derived from pooled data can be tested for significance, but investigators should keep in mind the limits on interpretation recognized by their field.

Third, and again when observations derive from different subjects, investigators may prefer to treat statistics (e.g., Yule's Q's) associated with different sequences just as scores to be analyzed using standard techniques like t test or the analysis of variance (see Wickens, 1993). In such cases, statistics for the sequences under consideration would be computed separately for each subject. However, analyses of these statistics tell us only whether they are systematically affected by some research factor. They do not tell us whether the statistics analyzed are themselves significant. In order to determine that, we could test individual z scores for significance, assuming enough data, and report for how many participants z scores were significant, or else compute a single z score from pooled data, assuming that pooling over units seems justified.

For example, Bakeman and Adamson (1984), for their study of infants' attention to people and objects, observed infants playing both with their mothers and with same-age peers. Coders segmented the stream of behavior into a number of mutually exclusive and exhaustive behavioral states: Two of those states were "Supported Joint" attention (infant and the other person were both paying attention to the same object) and "Object" attention (the infant alone was paying attention to some object). The Supported Joint state was not especially common when infants played with peers. For that reason, observations were pooled across infants, but separately for the "with mother" and "with peer" observations.

The z scores computed from the pooled data for the Supported Joint to Object transition were large and significant, both when infants were observed with mother and when with peers. This indicates that, considering these observations as a whole, the Supported Joint to Object sequence occurred significantly more often than expected, no matter the partner. In addition, an analysis of variance of individual scores indicated a significant partner effect, favoring the mother. Thus, not only was this sequence significantly more likely than expected with both mothers and peers, the extent to which it exceeded the expected was significantly higher with mothers, compared to peers.

In general, we suspect that most of our colleagues (and journal editors) are uneasy when data are pooled over human participants. Thus it may be worthwhile to consider how data such as those just described might be analyzed, not only avoiding pooling, but actually emphasizing individuality. The usual parametric tests analyze group means and so lose individuality. Better, it might be argued, to report how many subjects actually reflected a particular pattern, and then determine whether that pattern was observed in more subjects than one might expect by chance.

For such analyses, the simple sign test suffices. For example, we might report the number of subjects for whom the Yule's Q associated with the Supported Joint to Object transition was positive when observed with mothers, and again when observed with peers. If 28 infants were observed, by chance alone we would expect to see the pattern in 14 (50%) of them, but if the number of positive Yule's Q's was 19 or greater ($p < .05$, one-tailed sign test), we would conclude that the Supported Joint to Object transition was evidenced by significantly more infants than expected. And if the Yule's Q when infants were observed with mothers was greater than the Yule's Q when infants were observed with peers for 19 or more infants, we would conclude that the association was stronger when with mothers, compared to peers. The advantage of such a sign-test approach is that we learn, not just what the average pattern was, but exactly how many participants evidenced that pattern.

The approach presented earlier in section 8.2 can also be applied to the issue of pooling over subjects. Again, the question we ask is one of homogeneity, that is, whether the sequential structure is the same across subjects, or groups of subjects (instead of across time as for stationarity). The formula to test this possibility is similar to the formula for stationarity (see Gottman & Roy, 1990, pp. 67 ff.). In the following formula the sum is across $s = 1, 2, \ldots S$ subjects, and *P(IJ)* represents the pooled joint probability across the s subjects:

$$G^2 = 2 \sum_s N(IJ, s) \log_e \left(\frac{P(IJ, s)}{P(IJ)} \right)$$

The degrees of freedom are $(s - 1)$ (NCODES)^r (NCODES-1), where r is the order of the chain (in our case, with lag-1, $r = 1$), where NCODES = the number of codes. Note that this approach is quite general. For example, we can test whether a particular married couple's sequential data is best classified with one group of couples or with another group. Although tedious, this could be a strategy for grouping subjects.

8.5 How many data points are enough?

The issue of the number of tallies – of determining how many data points are enough – remains an important matter. The question the investigator needs to ask is this: How many events need to be coded in order to justify assigning significance to a computed z score associated with a particular cell or a chi-square statistic associated with a table?

At issue is not just the total number of events, but how they are distributed among the possible codes. It is worth reflecting for a moment why – and when – we need be concerned with sufficient tallies for the various codes in the first place. The bedrock reason is stability. If a summary statistic like a chi-square, a z score, or Yule's Q is based on few tallies, then quite rightly we place little confidence in the value computed, worrying that another time another observation might result in quite a different value.

For example, if one of the row sums of a 2×2 table is a very small number (such as 1 or 2), then shifting just one observation from one column to the other can result in a big change in the summary statistic. As a specific example, imagine that 50 observations were classified A|not-A and B|not-B, as follows:

	B	~ B	
A	2	0	2
~ A	24	24	48
	26	24	50

The Pearson chi-square for this table is 1.92 (so the z score is its square root, 1.39) and its Yule's Q is $+1$. After all, every A (all two of them) was followed by a B. However, if only one of the A's were followed by a B,

	B	~ B	
A	1	1	2
~ A	24	24	48
	25	25	50

then all statistics (Pearson chi-square, z score, and Yule's Q) would be zero. This example demonstrates summary statistics' instability when only a few instances of a critical code are observed.

To protect ourselves against this source of instability, we compute summary statistics only when all marginals sums are 5 or greater, and regard the value of the summary statistics as *missing* (too few instances to regard the computed values as accurate) in other cases. This is only an arbitrary rule, of course, and hardly confers absolute protection. Investigators should always be alert to the scanty data problem, and interpret results cautiously when summary statistics (e.g., z scores, Yule's Q's) are based on few instances of critical codes.

If stability is the bedrock reason to be concerned about the adequacy of the data collected, correct inference is a secondary but perhaps more often mentioned concern. This matters only when investigators wish to assign a p value to a summary statistic (e.g., a X^2 or a z score) based on assumptions that the statistic follows a known distribution (e.g., the chi-square or normal). Guidelines for amount of adequate data for inference have long been addressed in the chi-square and log-liner literature, so it makes sense to adapt those guidelines – many of which are stated in terms of expected frequencies – for lag-sequential analysis.

Several considerations play a role, so absolute guidelines are as difficult to define as they are desired. Summarizing current advice, Wickens (1989, p. 30) noted that (a) expected frequencies for two-dimensional tables with 1 degree of freedom should exceed 2 or 3 but that with more degrees of freedom some expected frequencies may be near 1 and with large tables up to 20% may be less than 1, (b) the total sample should be at least four or five times the number of cells (more if marginal categories are not equally likely), and (c) similar rules apply when testing whether a model fits a three- or larger-dimensional table.

As noted earlier, when lag 1 effects are studied, the number of cells is K^2 when consecutive codes may repeat and $K(K-1)$ when they cannot. Thus, at a minimum, the total number of tallies should exceed K^2 or $K(K-1)$, as appropriate, times 4 or 5. Additionally, marginals and expected frequencies should be checked to see whether they also meet the guidelines. When lag L effects are studied, the number of cells is K^{L+1} when consecutive codes may repeat and $K(K-1)^L$ when they cannot. As L increases, the product of these values multiplied by 4 or 5 can become discouragingly large. However, if attention is limited to three-dimensional $0(L - 1)L$ tables, as suggested in section 7.6, then the number of cells is no greater than K^3 when consecutive codes may repeat and $K^2(K - 1)$ when they cannot. But remember, these values multiplied by 4 or 5 represent a minimum number of tallies. Marginals and expected frequencies still need to be examined. Further, these products provide a guideline for the minimum number of events that should be coded.

Without question, considerable data may be required for a sequential or log-linear analysis. Following the strategy that limits attention to three-dimensional tables (section 7.6), K is the determining factor. For example, the numbers of cells when K is 3, 5, 7, and 9 and consecutive codes may repeat are

	L:1	2
K:3	9	27
5	25	125
7	49	343
9	81	729

(values for higher lags are the same as when $L = 2$; the general formula is K^2 when $L = 1$ and K^3 when $L = 2$ or higher) and when consecutive codes cannot repeat, numbers of cells (excluding structural zeros) are

	L:1	2	3
K:3	6	12	18
5	20	80	100
7	42	252	294
9	72	576	648

(values for higher lags are the same as when $L = 3$; the general formula is $K(K - 1)$ when $L = 1$, $K(K - 1)^2$ when $L = 2$, and $K^2(K - 1)$ when $L = 3$ or higher). Taking into account only the times-the-number-of-cells rule, in order to compute the N needed these values should be multiplied by 4, 5, or whatever factor seems justified (Bakeman & Quera, 1995b). Moreover, as Bakeman and Robinson (1994) note, such guidelines should not be regarded as minimal goals to be satisfied, but as troubled frontiers from which as much distance as possible is desired.

If the numbers still seem onerous, reconsider expected frequencies and note that Haberman (1977; cited in Wickens, 1989, p. 30) suggested that requirements for tests based on the difference between two chi-squares – such as the tests of $0 \perp L|L - 1$ described in section 7.6 – may depend more on frequency per degree of freedom than on the minimum expected cell size. When doubts arise, it may be best to consult local experts. But for many analyses, especially when the number of codes analyzed (K) is large, expect that the number of events coded should number in the thousands, not hundreds.

The guidelines described in the preceding paragraphs assumed that the computed test statistics would be distributed exactly as some theoretical distribution (e.g., the chi-square or normal), thus permitting p values to be based on the appropriate theoretic distribution. Such tests are often called *asymptotic* because as the amount of data on which the test statistic is based increases, its distribution approximates the theoretical one ever more closely. Asymptotic tests may be either parametric like those based on z or nonparametric like those based on chi-square. Permutation tests (Edgington, 1987; Good, 1994), although less well known, provide an alternative. Such tests construct sampling distributions from the data observed. No reference is made to another, theoretic distribution, so no minimum-data assumption to justify the reference is required.

Especially when data are few, investigators should consider permutation tests as an alternative to the usual asymptotic ones. They yield an exact, instead of an asymptotic, p value, and render minimum-data requirements unnecessary. If data are few, statistical significance will still be unlikely, but that is at it should be. Interested readers are urged to consult Bakeman, Robinson, and Quera (1996) for more details concerning permutation tests in a sequential context. But no matter whether asymptotic or permutation tests are used, you still should expect that the number of events coded will often need to number in the thousands.

8.6 The type I error problem

Even when enough data are collected to justify significance testing for the various scores computed, the problem of type I error – of claiming that sequences are "significant" when in fact they are not – remains. The reason type I error is such a problem with sequential analyses such as those described earlier is that typically investigators have many codes. These many codes generate many more possible sequences, especially when anything longer than two-event sequences is considered. The number of sequences tested for significance can rapidly become astronomical, in which case the probability of type I error (the alpha level) approaches certainty. When tests are independent, and the alpha level for each is .05, the "investigationwise" or "studywise" alpha level, when k tests are performed, is $1 - .95^k$ (Cohen & Cohen, 1983). Thus if 20 independent tests are performed, the probability of type I error is really .64, not .05.

What each study needs is a coherent plan for controlling type I error. The best way, of course, is to limit drastically the number of tests made. And even then, it makes sense to apply some technique that will assure a desired studywise alpha level. For example, if k tests are performed and a studywise alpha level of .05 is desired, then, using Bonferroni's correction, the alpha level applied to each test should not be alpha, but rather alpha divided by k (see Miller, 1966). Thus if 20 tests are performed, the alpha level for each one should be .0025 (.05/20).

When studies are confirmatory, type I error usually should not be a major problem. Presumably in such cases the investigator is interested in (and will test for significance) just a few theoretically relevant sequences. Exploratory studies are more problematic. Consider the parallel play study discussed earlier. Only five codes were used, which is not a great number at all. Yet these generate 20 possible two-event and 80 possible three-event sequences. This makes us think that unless very few codes are used (three or four, say) and unless there are compelling reasons to do so, most exploratory investigations should limit themselves to examining just two-event sequences, no longer – even if the amount of data is no problem.

Even when attention is confined to two-event sequences, the number of codes should likewise be limited. For two-event sequences, the number of possible sequences, and hence the number of tests, increase roughly as the square of the number of codes. For this reason, we think that coding schemes with more than 10 codes border on the unwieldy, at least when the aims of a study are essentially exploratory.

Two ways to control type I error were described in section 7.6 when discussing log-linear approaches to lag-sequential analysis. First, exploratory studies should not fish for effects at lag L in the absence of significant lag L omnibus tests. And second, the set of seemingly significant sequences at lag L should be winnowed into a smaller subset that can be viewed as responsible for the model of independence's failure to fit. Still, as emphasized in section 7.7 when discussing Yule's Q, guiding ideas provide the best protection against type 1 error. Investigators should always be alert for ways to limit the number of statistical tests in the first place.

8.7 Summary

Several issues important if not necessarily unique to sequential analysis have been discussed in this chapter. Investigators should always worry whether summary indices (means, Yule's Q's, etc.) are based on sufficient data. If not, confidence in computed values and their descriptive value is seriously compromised. Further, when inferential statistics are used, data sufficient to support their assumptions are required. Guidelines based on log-linear analyses were presented here, but the possibility of permutation tests, which require drastically fewer assumptions, was also mentioned. Again, investigators should always limit the number of statistical tests in any study, else they court type I error. Of help here is the discipline provided by guiding ideas and theories, clearly stated. In contrast, issues of pooling may arise more in sequential than other sorts of analyses because of data demands. Pooling data over units such as individuals, dyads, families, etc., is rarely recommended, no matter how necessary it seems. When data per unit are few, a jackknife technique (computing several values for a summary statistic, each with data for a different unit removed, then examining the distribution for coherence) is probably better than pooling. Finally, common to almost all statistical tests is the demand for independence (or exchangeability; see Good, 1994). When two-event chains are sampled in an overlapping manner from longer sequences, this requirement might seem violated, but simulation studies indicate that the apparent violation in this particular case does not seem consequential.

9

Analyzing time sequences

9.1 The tyranny of time

Events unfold in time. What distinguishes users of sequential analysis from many other researchers is that they attempt to move beyond this banal observation and try to capture aspects of this unfolding in quantifiable ways. For such purposes, it is often sufficient to know just the order of events, and to use the techniques for analyzing event sequences discussed in chapter 7.

Often, however, we want to know more than just the order of events. We want to know in addition how people (dyads, animals, etc.) spent their time. For that reason, it is common for investigators to record not just the *order* of events, but their *times* as well. In chapter 3 we described three ways such time information could be recorded (timing onsets and offsets, timing pattern changes, interval recording), and in chapter 5 we described three ways data could be represented preserving time (state sequences, timed-event sequences, interval sequences). Once recorded and represented, however, sometimes time information exerts a tyrannical hold on investigators who then seem reluctant to omit time from their analyses, even when this would be appropriate.

For many studies, especially when behavioral states are coded, we think time is worth recording primarily because "time-budget" information (amounts or percentages of time devoted to different kinds of events or behavioral states) has such descriptive value. For example, Bakeman and Adamson (1984), in their study of infants' attention to objects and people, recorded onset times for behavioral states, represented these data as state sequences, and computed and reported percentages of time devoted to the various behavioral states when with different partners (mothers or peers) at different ages.

However, when examining how behavioral states were sequenced, Bakeman and Adamson ignored time, in effect reducing their state-sequential data to event sequences. This approach has merit whenever investigators want to describe how a simple stream of events or states unfolded in time. For example, if state sequences are analyzed instead of event sequences, using a time unit instead of the event as the basic unit of analysis, values for transitional probabilities are affected by how long particular events lasted – which is undesirable if all the investigator wants to do is describe the typical sequencing of events.

As an example, recall the study of parallel play introduced earlier, which used an interval recording strategy (intervals were 15 seconds). Let U = unoccupied, S = solitary, and P = Person Play. Then:

Interval = 15; U, U, U, S, P, P . . .

Interval = 15; U, S, S, S, P, P . . .

Interval = 15; U, U, S, S, P, P . . .

represent three slightly different ways an observational session might have begun. All three interval sequences clearly represent one event sequence: Unoccupied to Solitary to Parallel. Yet values for transitional probabilities and their associated z scores would be quite different for these three interval sequences. For example, the $p(P_{t+1}|S_t)$ would vary from 1.00 to 0.33 to 0.50 for the three sequences given above. Worse, the z score associated with 0.33 would be negative, whereas the other two z scores would be positive. No one would actually compute values for sequences of just six intervals, of course, but if we had analyzed longer interval sequences like these with the techniques described in chapter 7, it is not clear that the USP pattern would have been revealed. Very likely, especially if each interval had represented 1 second instead of 5 seconds, we might have discovered only that transitions from one code to itself were likely, whereas all other transitions were unlikely. In short, we urge investigators to resist the tyranny of time. Even when time information has been recorded, it should be "squeezed out" of the data whenever describing the typical sequencing of events is the primary concern. In fact, the GSEQ program (Bakeman & Quera, 1995a) includes a command that removes time information, thereby transforming state, timed-event, or interval sequences into event sequences when such is desired.

9.2 Taking time into account

The simple example of a USP sequence presented in the previous section hints at the underlying unity of the four methods of representing data described in chapter 5. This unity, in turn, suggests a welcome economy. If we have already learned a number of techniques for analyzing event sequences (in chapter 7), and if state, timed-event, and interval sequences are logically the same as event sequences, then analyzing time sequences (sequential data that take time into account, i.e., state, timed-event, and interval sequences) does not require learning new techniques, only the application of old ones.

In fact, one underlying format suffices for event, state, timed-event, and interval sequences. These four forms are treated separately both here and in the Sequential Data Interchange Standard (SDIS; see Bakeman & Quera, 1992) because this connects most easily with what investigators actually do, and have done historically. Thus the four forms facilitate human use and learning. A general-purpose computer program like GSEQ, however, is better served by a common underlying format because this allows for greater generality and hence less specific-purpose computer code. Indeed, the SDIS program converts SDS files (files containing data that follow SDIS conventions) into a common format (called MDS or modified SDS files) that is easily read by GSEQ (Bakeman & Quera, 1995a).

The technical details need not concern users of these computer programs, but understanding the conceptual unity of the four forms can be useful. Common to all four is an underlying metric. For event sequences, the underlying metric is the discrete event itself. For state and timed-event sequences, the underlying metric is a unit of time, often a second. And for interval sequences, the underlying metric is a discrete interval, usually (but not necessarily) defined in terms of time.

The metric can be imagined as cross marks on a time line, where the space between cross marks is thought of as bins to which codes may be assigned, each representing the appropriate unit. For event sequences, one code and one code only is placed in each bin. Sometimes adjacent bins may be assigned the same code (consecutive codes may repeat), sometimes not (for logical reasons, consecutive codes cannot repeat). For state sequences, one (single stream) or more codes (multiple streams) may be placed in each bin. Depending on the time unit used and the typical duration of a state, often a stretch of successive bins will contain the same code. For timed-event sequences, one or more codes or no codes at all may be placed in each bin. And for interval sequences, again one or more codes or no codes at all may be placed in each bin. As you can see, the underlying structure of all forms is alike. Successive bins represent successive units and, depending on the form, may contain one or more or no codes at all.

Interval sequences, in particular, can be quite useful, even when data were not interval recorded in the first place. For example, imagine that a series of interactive episodes are observed for particular children and that attributes of each episode are recorded (e.g., the partner involved, the antecedent circumstance, the type of the interaction, the outcome). Here the *event* (or episode) is multidimensional, so the event sequential form is not adequate. But interval sequences, which permit several codes per bin, work well, and permit both concurrent (e.g., are certain antecedents often linked with particular types of interaction) and sequential (e.g., are consequences of successive episodes linked in any way) analyses. Used in this way, each episode defines an interval (instead of some period of elapsed time); in such cases, interval sequences might better be called multidimensional events. Further examples of the creative and flexible use of the four forms for representing sequential data are given in Bakeman and Quera (1995a, especially chapter 10).

Because the underlying form is the same for these four ways of representing sequential data, computational and analytic techniques are essentially the same (primarily those described in chapter 7). New techniques need not be introduced when time is taken into account. Only interpretation varies, depending on the unit, whether an event or a time unit. For event sequences, coders make a decision (i.e., decide which code to assign) for each event. For interval sequences, coders make decisions (i.e., decide which codes occurred) for each interval. For state and timed-event sequences, there is no simple one-to-one correspondence between decisions and units. Coders decide and record when events or states began and ended. They may note the onset of a particular event, the moment it occurs. But just as Charles Babbage questioned the literal accuracy of Alfred Lord Tennyson's couplet "Every minute dies a man, / Every minute one is born" (Morrison & Morrison, 1961), so too we should question the literal accuracy of a claim that observers record onsets discretely second by second and recognize the fairly arbitrary nature of time units. For example, we can double tallies by changing units from 1 to 1/2 second. The connection, or lack of connection, between coders' decisions and representational units is important to keep in mind and emphasize when interpreting sequential results.

9.3 Micro to macro

In this section we would like to share some wisdom based on our experience doing programmatic observational research with the same kinds of data for over a decade. Sequential analysis is interesting because so much theoretical clarity about interacting people is provided by the study of temporal patterns. Often when we have begun working in an area, we start with fairly small units and a large catalog of precise codes. A microanalytic description often is the product of these initial efforts (e.g., Brown, Bakeman, Snyder, Fredrickson, Morgan, & Hepler, 1975).

Sequential analysis of the micro codes then can be used to identify indexes of more complex social processes (e.g., Bakeman & Brown, 1977). For example, Gottman (1983) found that a child's clarification of a message after a request for clarification (Hand me the truck/ Which truck?/ The red truck) was an index of how connected and dialogic the children's conversations were. This sequence thus indexed a more macro social process. Gottman could have proceeded to analyze longer sequences in a statistical fashion, but with 40 codes the four-element chain matrix will contain 2,560,000 cells! Most of these cells would have been empty, of course, but the task of even looking at this matrix is overwhelming. Instead, Gottman designed a macro coding system whose task it was to *code* for sequences, to code larger social processes. The macro system used a larger interaction unit, and it gave fewer data for each conversation (i.e., fewer units of observation). However, the macro system used a larger interaction unit, and it gave fewer data for each conversation (i.e., fewer units of observation). However, the macro system was extremely useful. First, it was far more rapid to use than the micro system. Second, because a larger unit was now being used, new kinds of sequences were being discovered. This revealed an organization of the conversations that Gottman did not notice, even with the sequential analysis of the micro data (see Gottman & Parker, 1985 in press).

To summarize, one strategy we recommend for sequential analysis is *not* looking for everything that is patterned by employing statistical analysis of one data set. It is possible to return to the data, armed with a knowledge of patterns, and to reexamine the data for larger organizational units.

9.4 Time-series analysis

Time-series analysis offers a wide range of analytic options (see Gottman, 1981, for a comprehensive introduction), and, furthermore, it is possible to create time-series data from a categorial stream of codes, as we mentioned in section 5.7. In this section we shall review a few of the advantages provided by time-series analysis, and discuss further how to create time-series data from a stream of categorical data.

What are the advantages of creating time-series data from a stream of categorial data? First, one can observe the overall shape of the interaction. This can be useful in two ways. First, one can study economies of interactions. For example, Gottman (1979) created a time-series from data obtained from marital interaction. The value of the variable was the total positive minus the total negative codes up to that point in time. In some couples, both the husband's and wife's graphs were quite negative. These couples tended to be high in reciprocating negative affect. In some couples, one partner's graph was negative and the other's was positive. These couples tended to have one partner who gave in most of the time in response to the partner's complaints. Couples whose graphs were flat at

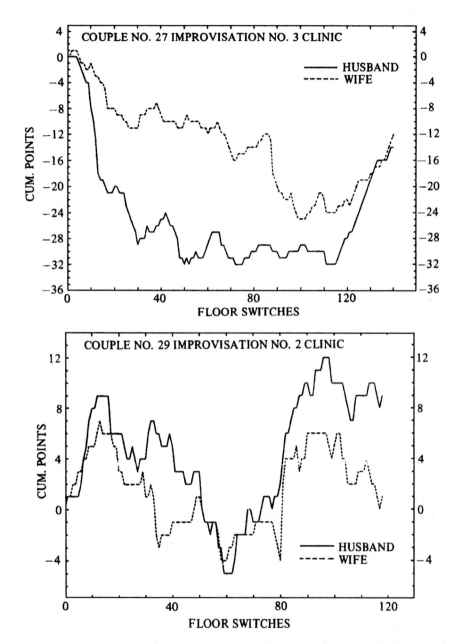

Figure 9.1. Time series for two clinic couples. From Gottman (1979b, p. 215).

the start of an interaction tended to have social skill deficits in listening, whereas couples whose graphs were flat at the end of an interaction tended to have social skill deficits in negotiation.

A second use of a time-series graph is that it makes it possible to *discover the limitations of a coding system.* One can use the graphs to scan for shifts in slope or level. These may be rare critical events that the coding system itself does not know about. An example of this comes from a videotape that Ted Jacob had which was coded using the Marital Interaction Coding System (MICS). The interaction began in a very negative way but changed dramatically in the middle and became quite positive. The event that triggered the change seemed to be the husband's summarizing what he thought was the wife's complaint and then accepting responsibility for the problem. The MICS has no code for summarizing the other (which is a very rare event, but quite powerful); it does have a code for accepting responsibility, but this was miscoded on this tape. Despite the fact that the critical event was missed by the coding system, time-series analysis of the data detected the shift in the overall positivity of the interaction and pinpointed the time of the switch. Gottman refers to this use of time series as "Gödeling" because, like Kurt Gödel's work, it is concerned with using a system to view itself and discover its own limitations.

Creating time series from categorical data

In addition to these reasons for creating time-series data from categorical data, time-series analysis has some powerful analytic options, which we shall discuss briefly in a moment. First we would like to mention three options for creating time-series data from categorical data.

One option, used by psychophysiologists, is the *interevent interval.* This involves graphing the time between events of a certain type over time. Cardiovascular psychophysiologists, for example, plot the average time between heart beats within a time block of sufficient size. Interevent intervals can be computed with quite simple data. For example, imagine we had asked a smoker to keep a diary, noting each time he or she smoked a cigarette, and that a small portion of the data looks like that portrayed in Table 9.1. (For convenience, all times are given to the nearest 6 minutes or 0.1 hour.) Such interevent intervals are often averaged in some way. For example, if we used 2-hour time blocks, the averaged interevent intervals would be as given in Table 9.1. These scores are then available for graphing or subsequent time-series analysis.

A second option is the *moving probability-window.* Here we compute the proportion of times an event was observed within a particular block of

Table 9.1. *Computing interevent interval time series*

Time	Interevent interval	Block	Block average
7:54 a.m.			
----------------------------------	2:12 (2.2)	----------------------	
10:06			
	0:42 (0.7)	10–12	1.45
10:48			
----------------------------------	1:30 (1.5)	----------------------	
12:18			
	0:24 (0.4)	12–2	0.95
12:42			
----------------------------------	1:48 (1.8)	----------------------	
2:30			
	1:18 (1.8)	2–4	1.55
3:48			

Note: Interevent intervals are given both in hours : minutes and in decimal hours.

observations, and then we slide the window forward in time. This option *smooths* the data as we use a larger and larger time unit, a useful procedure for graphical display, but not necessary for many time-series procedures (particularly those in the frequency domain).

A third option is the *univariate scaling* of codes. For two different approaches to this, see Brazelton, Koslowski, and Main (1974), and Gottman (1979b).

Each option produces a set of time series for each variable created, for each person in the interacting unit. Analysis proceeds within each interacting unit ("subject"), and statistics of sequential connection are then extracted for standard analysis of variance or regression. A detailed example of the analysis of this kind of data obtained from mother–infant interaction appears in Gottman, Rose, and Mettetal (1982). A review of time-series techniques is available in Gottman's (1981) book, together with 10 computer programs (Williams & Gottman, 1981).

Brief overview of time-series analysis

Brazelton, Koslowski, and Main (1974) wrote about the interactive cyclicity and rhythmicity that they believe characterizes the face-to-face play of mother and infant. They described a cycle of attention followed by the withdrawal of attention, with each partner waiting for a response from the other. They described the interactional energy building up, then ebbing and cycling in synchrony. Part of this analysis had to do with their confidence in the validity of their time-series variable. They wrote:

> In other words, the strength of the dyadic interaction dominates the meaning of each member's behavior. The behavior of any one member becomes a part of a cluster of behaviors which interact with a cluster of behaviors from the other member of the dyad. No single behavior can be separated from the cluster for analysis without losing its meaning in the sequence. The effect of clustering and of sequencing takes over in assessing the value of particular behaviors, and in the same way the dyadic nature of interaction supercedes the importance of an individual member's clusters and sequences. (p. 56)

Time-series analysis is ideally suited to quantitative assessments of such descriptions. It is a branch of mathematics that began to be developed in the mid-1700s, on the basis of a suggestion by Daniel Bernoulli, and later developed into a theorem by Jean Baptiste Joseph Fourier in 1822. The idea was that any continuous function could be best approximated in a least-squares sense by a set of sine and cosine functions. At first this idea seemed counterintuitive to mathematicians. However, it is true, and it is true even if the function itself is not periodic. Sines and cosines have an advantage over polynomials, because polynomials tend to wander off to plus or minus infinity after the approximation period, which is very poor for extrapolation if one believes that the data have any repetitive pattern. This is usually the case in our scientific work. For example, we tend to believe that if we observe the brightness fluctuations of a variable star, it really doesn't matter very much if we begin observing on a Monday or a Tuesday; we believe that the same process is responsible for generating the data, and that it has a continuity or stability (which, in time-series language is referred to as "stationarity").

Fourier proved his famous theorem incorrectly, and proving it correctly took the best mathematical minds over a century; furthermore, it has led to the development of much of a branch of modern mathematics called "analysis." Time-series analysis underwent a major conceptual revolution in the 20th century because of the thinking of an American, Wiener, and a Russian, Khintchine.

Without time-series analysis, attempts to describe cyclicity and synchronicity end in a hopeless muddle of poetic metaphor about interactive patterns. For example, Condon and Ogston (1967) tried to summarize 15 minutes of the dinner time interaction of one family. They wrote:

> We are dealing with ordered patterns of change during change, which exhibit rhythmic and varying patterns in the temporal sequencing of such changes. Metaphorically, there are waves within waves within waves, with complex yet

determinable relationships between the peaks and troughs of the levels of waves, which serve to express organized processes with continually changing relationships. (p. 224)

Time-series analysis offers us a well-elaborated set of procedures for passing beyond metaphor to a precise analysis of cyclicity, synchronicity, and the analysis of the relationship between two time series (called "cross correlation"), controlling for regularity within each time series (called "autocorrelation"), among other options.

The notion of cycles

A good way to begin thinking of cycles is to imagine a pendulum moving back and forth. The amplitude of the oscillation of the pendulum is related to the energy with which it is shoved. In fact, the variance of the pendulum's motion is proportional to this energy, and this is proportional to the square of the amplitude. The period of oscillation of the pendulum is the time it takes for the pendulum to return to the same spot; it is usually measured as the time from peak to peak of oscillations. Now imagine that we attach a second pendulum to the first and permit them to be able to oscillate independently. We can generate very complex oscillations just with the oscillations of two pendula. A new variable enters into the picture when we imagine two pendula, the relative phase of oscillation of the two. They can be moving in synchrony, or exactly opposite (180 degrees out of phase), or somewhere in between. So now we have three parameters: the energy of each pendulum (proportional to the amplitude squared), the frequency of oscillation of the pendulum (which is the reciprocal of the period of oscillation), and the relative phases of the pendula. These are the basic dimensions of what is called "frequency domain" time-series analysis.

Intuitive motions of the spectrum

What is the spectrum? Imagine Isaac Newton holding a prism through which white light passes on one side and the rainbow emerges from the other. The rainbow is called the spectrum; in general, it is the resolution of some incoming wave into its basic components. In the case of white light, all colors (or frequencies) are present in all brightnesses (energies, variances). For different kinds of oscillations, some of the colors would be missing entirely, some would be weaker, some would be stronger. Furthermore, the phase relationships of the different colors could vary. The

Table 9.2. *Guessing that the period t = 5*

	1	2	3	4	5 ·
	0.00	0.95	0.59	−0.59	−0.95
	0.00	0.95	0.59	−0.59	−0.95
	0.00	0.95	0.59	−0.59	−0.95
	0.00	0.95	0.59	−0.59	−0.95
	0.00	0.95	0.59	−0.59	−0.95
	0.00	0.95	0.59	−0.59	−0.95
Means	0.00	0.95	0.59	−0.59	−0.95

Table 9.3. *Guessing incorrectly that the period t = 4*

	1	2	3	4
	0.00	0.95	0.59	−0.59
	−0.95	0.00	0.95	0.59
	−0.59	−0.95	0.00	0.95
	0.59	−0.59	−0.95	0.00
	0.95	0.59	−0.59	−0.99
Means	0.00	0.00	0.00	0.00

spectrum, or the "spectral density function," tells us only which frequencies are present and to which degree; that is, we learn how much variance each frequency accounts for in the given time series.

We shall illustrate how this spectral density function is computed by using an old 19th-century method, called the periodogram (a technique no longer used). Suppose we generate a very simple time series, $x_t = \sin(1.257t)$, and let t go from 0 to 10. The values of the time series are 0.00, 0.95, 0.59, −0.59, −0.95, 0.00, 0.95, 0.59, −0.59, −0.95, 0.00. Suppose we did not know that these data repeated every five time intervals. Suppose we guess at the period and guess correctly that the period is 5, and we arrange the data as shown (Table 9.2). In this table, we can see that the variance of the series is equal to the variance of the means. The ratio of these two variances is always one when we have guessed the right period. Suppose we had guessed the wrong period, say $t = 4$. This situation is illustrated in Table 9.3.

In the case of Table 9.3, the variance of the means is zero, so that the ratio of the variance of the means to the variance of the series is zero. If we were to plot the value of this ratio for every frequency we guess (remember frequency equals $1/t$), this graph is an estimate of the spectral density function. Of course, this is the ideal case. In practice, there would be a lot of noise in the data, so that the zero values would be nonzero, and also the peak of the spectral density function would not be such a sharp spike. This latter modification in thinking, in which the amplitudes of the cycles are themselves random variables, is a major conceptual revolution in thinking about data over time; it is the contribution of the 20th century to this area. (For more discussion, see Gottman, 1981.)

Rare events

One of the uses of *univariate* time-series analysis is in evaluating the effects of rare events. It is nearly impossible to assess the effect of a rare but theoretically important event without pooling data across subjects in a study by the use of sequential analysis of categorical data. However, if we create a time-series variable that can serve as an *index* of the interaction, the problem can be solved by the use of the interrupted, time-series quasi-experiment.

What we mean by an "index" variable is one that is a meaningful theoretical index of how the interaction is going. Brazelton, Koslowski, and Main (1974) suggested an index time series that measured a dimension of engagement and involvement to disengagement. The dimension assessed the amount of interactive energy and involvement that a mother expressed toward her baby and that the baby expressed toward the mother. This is an example of such an index variable. Gottman (1979) created a time-series variable that was the cumulative positive-minus-negative affect in a marital interaction for husband and wife. The interactive unit was the two-turn unit, called a "floor switch." Figure 9.1 illustrates the Gottman index time series for two couples.

Now suppose that the data in Figure 9.2 represented precisely such a point graph of a wife's negative affect, and that a rare but interesting event occurred at time 30, when her husband referred to a previous relationship he had had. We want to know whether this event had any impact on the interaction. To answer this question, we can use an interrupted time-series analysis. There are many ways to do this analysis (see Gottman, 1981). We analyzed these data with the Gottman–Williams program ITSE (see Williams & Gottman, 1981) and found that there was no significant change in the slope of the series [$t(32) = -0.3$], but that there was a significant effect in the change in level of the series [$t(32) = 5.4$]. This is one important use of time-series analysis.

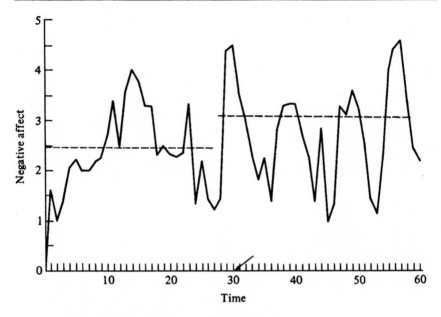

Figure 9.2. Interrupted time-series experiment.

Cyclicity

Univariate time-series analysis can also answer questions about the *cyclicity* of the data. Recall that to assess the cyclicity of the data, a function called the "spectral density function" is computed. This function will have a significant peak for cycles that account for major amounts of variance in the series, relative to what we might have expected if there were no significant cycles in the data (i.e., the data were noise). We used the data in Figure 9.2 for this analysis, and used the Gottman–Williams program SPEC. The output of SPEC is relatively easy to interpret. The solid line in Figure 9.3 is the program's estimate of the spectral density function, and the dotted line above and below the solid line is the 0.95 confidence interval. If the entire confidence interval is above the horizontal dashed line, the cycle is statistically significant. The x axis is a little unfamiliar to most readers, because it refers to "frequency," which in time-series analysis means cycles per time unit. It is the reciprocal of the period of oscillation. The peak cycle is at a frequency of 0.102, which corresponds to a period of 9.804 time periods. The data are cyclic indeed. It is important to realize that cyclicity in modern time-series analysis is a statistical concept. What we mean by this is that the period of oscillation is itself a random variable, with a distribution. Thus, the data in Figure 9.3 are *almost* periodic, not *precisely* periodic. Most phenomena in nature are actually of this sort.

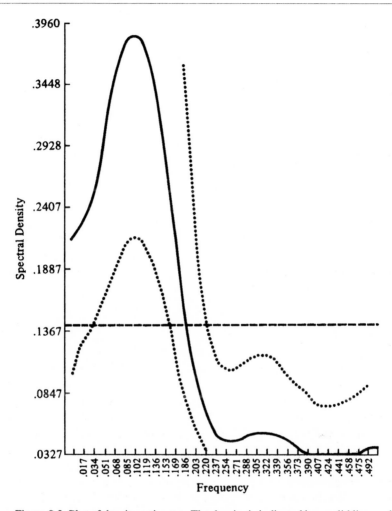

Figure 9.3. Plot of density estimates. The density is indicated by a solid line, and the 0.95 confidence interval by a dotted line. The white-noise spectrum is shown by a dashed line.

Multivariate time-series analysis

There are quite a few options for the multivariate analysis of time-series data. We shall discuss only one option here, a bivariate time-series analysis that controls for autocorrelation (predictability within each time series) in making inferences about cross correlation between two time series. This option is discussed in a paper by Gottman and Ringland (1981).

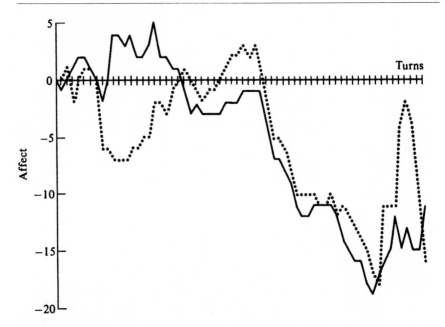

Figure 9.4. Couple mp48 conflict discussion. The solid line indicates the husband's affect; the broken line, the wife's affect.

The data in Figure 9.4 are a summary of the interactions of a happily
married couple discussing a marital conflict. The y axis is a cumulative
sum of positive minus negative affect in the discussion. This graph is fairly
typical of happily married couples' discussions of an area of continued
disagreement. The central third of the conversation shows a downward drift
because most disagreements occur in the middle phase of the discussion;
the final third shows the graph coming back up as the couple moves closer
to resolving the issue. In making an inference of influence from one partner
to another, what this procedure does is to attempt to predict as well as it
can from the parts of each series and then see if any additional information
is provided by the other series.

Table 9.4 summarizes the results of this analysis, which was accomplished with the program BIVAR from the Gottman–Williams package
(see Williams & Gottman, 1981). Row 1 of the table shows that the initial
starting values for the models is 8; this means that we shall go back 8 units
into the past; this number is arbitrary, but should probably not exceed 10
(it is limited by the total number of observations). The second row shows
results of the program's successive iterations to find the smallest model that
loses no information; this is the model with one autoregressive term for the
husband and four cross-regressive terms for the wife. The test of model 1

Table 9.4. *Bivariate-time series analysis of couple mp48's data*

Model	Husband	Wife	SSE	T LN(SSE/T)[a]
1	8	8	133.581	44.499
2	1	4	150.232	52.957
3	1	0	162.940	58.803

1 vs 2:	$Q = 8.458$	$df = 11$	$z = -.542$	
2 vs 3:	$Q = 5.846$	$df = 4$	$z = .653$	

Model	Husband	Wife	SSE	T LN(SSE/T)
1	8	8	207.735	76.291
2	5	2	224.135	81.762
3	5	0	298.577	102.410

1 vs 2:	$Q = 5.471$	$df = 9$	$z = -.832$	
2 vs 3:	$Q = 20.648$	$df = 2$	$z = 9.324$	

[a] Weighted error variance; see Gottman and Ringland (1981), p. 411.

versus model 2 should be nonsignificant, which is the case. The third row of the table shows the model with all of the wife's cross-regressive terms dropped. If this model is not significantly different from model 2, then we cannot conclude that the wife influences the husband; we can see that the z score (0.653) is not significant. Similar analyses appear in the second half of the table for determining if the husband influences the wife; the comparison of models 2 and 3 shows a highly significant z score (this is the normal approximation to the chi-square, not to be confused with the z score for sequential connection that we have been discussing). Hence we can conclude that the husband influences the wife, and this would be classified as a husband-dominant interaction according to Gottman's (1979) definition. We note in passing that although these time series are not stationary, it is still sensible to employ BIVAR.

Interevent Interval

In the study of the heart, in analyzing the electrocardiogram (ECG), there is a concept called "interbeat interval," or the "IBI." It is the time between the large R-spike of the ECG that signals the contraction of the ventricles of the heart. It is usually measured in milliseconds in humans. So, for example, if we recorded the ECG of a husband and wife talking to each other about a major issue in their marriage, and we took the average of the IBIs for each second (this can be weighted or prorated by how much of the second each IBI took up), we would get a time series that was a set of consecutive IBIs (in milliseconds) that looked like this: 650, 750, 635, 700, 600, 625, 704, and so on.

We wish to generalize this idea of IBI and discuss a concept we call the "interevent interval." This is like an old concept in psychology, the intertrial interval. When we are recording time, which we get for free with almost all computer-assisted coding systems, we also can compute the time between salient events, and these times can become a time series. When we do this, we are interested in how these interevent times change with time within an interaction.

Other ways of transforming observational data to a time series

When the observational data consist of ratings, the data are already in the form of a time series. For example, if we coded the amount of positive affect in the voice, face, body, and speech of a mother during parent – child interaction, we can add the ratings on each channel and obtain an overall time series.

When the data are categorical codes

There are other ways to obtain a time series from a stream of categorical codes. One method is the idea of a "moving probability window," which would be a window of a particular sized time block (say 30 seconds) within which we compute the frequencies of our codes (estimating their probabilities); then the window moves forward one time unit, and we compute these probabilities again. Another approach we have used (Gottman, 1979b; Gottman & Levenson, 1992), as has Tronick, Als, and Brazelton (1977; 1980) is to weight the codes along some composite dimension. In the Tronick case the dimension was a combination of engagement/disengagement and positive/negative, so that positive scores meant engaged and/or with positive affect, and negative scores meant disengaged and/or with negative affect. In the Gottman cases, positive and negative codes were given positive or negative integer weights, and the number of total positive minus negative points was computed for each turn at speech (a turn lasts about 6 seconds on marital interaction). We have found it very fruitful for visual inspection of a whole interaction to cumulate these time series as the interaction proceeds. Figure 9.5 (a) and (b) shows these time series *cumulated* to illustrate two very different marital interactions. Losada, Sanchez and Noble (1990), in their research on six-person executive work groups, code power and affect on 3- point scales and then turn these into numerical scores within time blocks. They then compute directional cross-correlations between people, and use these numbers to create an animated graphic of a "moving sociometric" of affect and power over time. This moving sociometric is time-locked so that the researchers can provide instant replay video feedback to the group so that the group can see a video illustration of their moving sociometric.

Why transform the data to a time series?

Reducing the observational coding to these summary time-series graphs is very profitable. In marital interaction having these time series made it possible for us (see Cook et al., 1995) to construct a mathematical model of the data that led us to a new theoretical language for describing our divorce prediction data and led us to a new discovery about what predicts divorce.

9.5 Autocorrelation and time-series analysis

Autocorrelation function

We will begin by examining the nature of the dependent time-structure of time-series data. To accomplish this we start by examining what is called the "autocorrelational structure" of the time series. This gives us information about how predictable the present data are from its past. To explain the concept of "lag-1" autocorrelation, we draw a lag-1 scatterplot of the data, where the x-axis is the data at time t, and the y-axis is the data at time $t + 1$. This means that we plot the data in pairs. The first point we plot is $(x1, x2)$; the second point is $(x2, x3)$; the third point is $(x3, x4)$; and so on. This gives a scatterplot similar to the ones we plot when we compute a regression line between two variables. The correlation in this case is called the "lag-1 autocorrelation coefficient," r_1. In a similar fashion we can pair points separated by two time points. The pairs of points would then be $(x1, x3)$, $(x2, x4)$, $(x3, x5)$, . . .; the two axes are $x(t)$ and $x(t + 2)$, and the autocorrelation coefficient would be the "lag-2 autocorrelation coefficient," r_2. We can continue in this fashion, plotting r_k, the lag-k autocorrelation coefficient against the lag, k. The autocorrelation function is useful in identifying the time-series model.

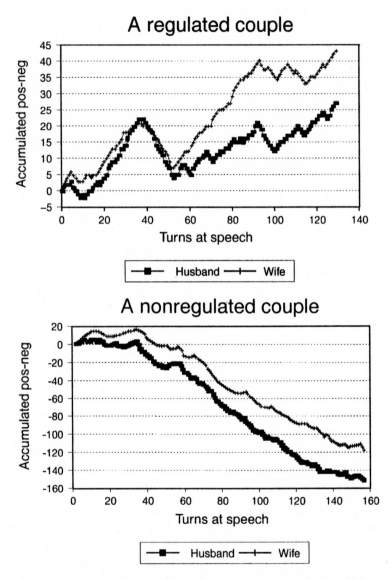

Figure 9.5. (a) Cumulative point graphs for a regulated couple, for which positive codes generally exceed negative codes. (b) Cumulative point graphs for a nonregulated couple, for which negative codes generally exceed positive codes.

Model identification

Using the autocorrelation function, an autoregressive model for the time series can be identified exactly if the series is stationary (this means that the series has the same correlation structure throughout and no local or global trends) using a computer program (see the Williams & Gottman, 1981, computer programs). A wide variety of patterns can be fit using the autoregressive models, including time series with one or many cycles.

Spectral time-series analysis

In 1822 Jean Baptiste Fourier, a French mathematician, discovered a very important theorem (one that took mathematicians over a hundred years to prove correctly and led to the development of a branch of modern mathematics called mathematical analysis). He discovered that any piecewise continuous function could be fit best (in the least-squares sense of distance) by a series of specially selected sine and cosine functions. These functions are, of course, cyclic, but the representation is still the best one ever if the function being fit is not itself cyclic. This was quite an amazing theorem.

In time-series analysis we can actually use the data to compute the cycles present in the data, even if these cycles are obscured with noise. This analysis is called a Fourier analysis, or a "spectral time-series analysis." In a spectral time-series analysis, the end result is that we generate a plot, called the "spectral density function," of the amount of variance accounted for by each of a set of cycles, from slow to fast. We do not use all cycles, but only a particular set, called "the overtone series." If a particular time series were actually composed of two cycles, a slow one and a fast one, the spectral density function would have peaks at these two cycle frequencies. Usually, however, real data do not look as spikelike as this figure, but instead the spectral density function is statistically significant across a band of frequencies.

We can then actually use these frequencies to fit a function to the data and use it to model the time series. In most cases this type of modeling is not feasible, because the models we obtain this way tend to be poor fits of the data. For this reason we usually use models like the autoregressive model.

Interrupted time-series experiments

Once we have a time series, and we have established that the time series itself has some validity (e.g., Gottman and Levenson's time series predicted divorce versus marital stability), we can model the series, and then we can scan the time series for statistically significant changes in overall level or slope. This is called an "interrupted time-series experiment," or ITSE (see Figure 9.6). An ITSE consists of a series of data points before and after an event generally called the "experiment." The "experiment" can be some naturally occurring event, in which case it is actually a quasi-experiment. We then represent the data before the intervention as one function $b_1 + m_1 t +$ Autoregressive term, and the data after the intervention as $b_2 + m_2 t +$ Autoregressive term. We need only supply the data and the order of the autoregressive terms we select, and the computer program tests for statistically significant changes in intercept (the b's) and slope (the m's). The experiment can last 1 second, or just for one time unit, or it can last longer. One useful procedure is to use the occurrence of individual codes as the event for the interrupted time-series experiment. Thus, we may ask questions such as "Does the wife's validation of her husband's upset change how he rates her?" Then we can do an interrupted time-series experiment for every occurrence of Wife Validation. For the data in Figure 9.6, there were 48 points before and 45 points after the validation. The order of the autoregressive term selected was about one tenth of the preintervention data, or 5. The t for change in intercept was $t(79) = -2.58$, $p < .01$, and the t for change in level was $t(79) = -1.70$, $p < .05$.

For this example, we used the Williams and Gottman (1981) computer program ITSE to test the statistical significance of changes in intercept and slope before and after the experiment; an autoregressive model of any order can be fit to the data. Recently, Crosbie (1995) developed a powerful new method for analyzing short time-series experiments. In these analyses only the first-order autoregressive parameter is used, and the preexperiment data are fit with one straight line (intercept and slope) and the postexperiment data are fit with a different straight line (intercept and slope). An omnibus F test and t tests for changes in level and slope are then computed. This method can be used to determine which codes in the observational system have potentially powerful impact on the overall quality of the interaction.

Phase-space plots

Another useful way to display time-series data is by using a "phase-space" plot. In a phase-space plot, which has an x-axis and a y-axis, we plot the data as a set of pairs of points: $(x1, x2), (x2, x3), (x3, x4), \ldots$ The x-axis is $x(t)$, and the y-axis is $x(t + 1)$, where t is time, $t = 1, 2, 3, 4$, and so on. Alternatively, if we are studying marital interaction, we can plot the interevent intervals for both husband and wife separately, so we have both

Husband rates wife

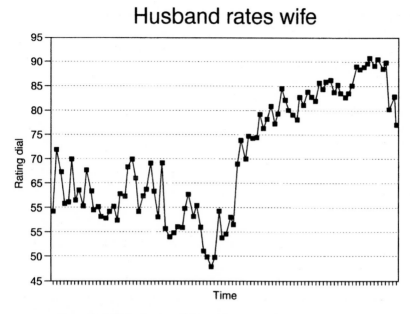

Figure 9.6. Plot of rating dial in which the husband rated his wife's affect during their conversation.

a husband and a wife time series. A real example may help clarify how we might use this idea (Gottman, 1990).

An example from an actual couple who subsequently divorced

For marital interaction many possible behavioral candidates exist for "the event" selected to be used for computing interevent intervals. As we have suggested, one promising event to select is negative affect, in part because it has a reasonably high base rate during conflict discussions, and because it tends to be high in dissatisfied marriages. Our interest here in computing the phase-space plot is not whether negative affect is high among an unhappily married couple, but whether this system is homeostatically regulated and stable, or whether it is chaotic. In this interaction of a married couple we computed the time between negative affect, independent of who it was (husband or wife) who displayed the negative affect. The times between negative affects were thus selected for analysis. These were interevent intervals for either partner in a marital interaction (Figure 9.7).

In phase-space plot the data look a lot like the scatterplot for the first-order autocorrelation coefficient, except for one thing. In the phase-space

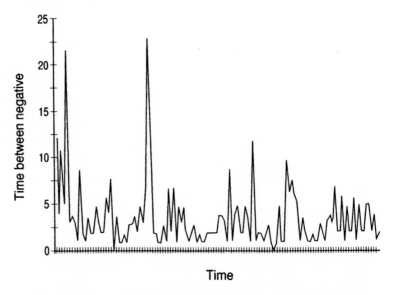

Figure 9.7. Interevent integrals for negative affect in a marital conversation.

plot we connect the successive dots with straight lines (see Figures 9.8 and 9.9). This gives us an idea of the "flow" of the data over time.

What does this figure mean in terms of the actual behavior of the marital system? It means that, insofar as we can ascertain from these data, we have a system whose energy balance is not stable, but dissipative; that is, it runs down. Like the pendulum winding down, this system tends toward what is called an attractor; in this case the attractor represents an interevent interval of zero. However, for the consequences of energy balance, this movement toward an attractor of zero interevent interval between negative affect may be disastrous. Specifically, this system tends, over time, toward shorter response times for negative affect. Think of what that means. As the couple talk, the times between negativity become shorter and shorter. This interaction is like a tennis match where that ball (negative affect) is getting hit back and returned faster and faster as the game proceeds. Eventually the system is tending toward uninterrupted negativity.

We can verify this description of this marital interaction using more standard analytic tools, in this case by performing the mathematical procedure we discussed called spectral time-series analysis of these IEIs (see Gottman, 1981). Recall that a spectral time-series analysis tells us whether there are specific cyclicities in the data, and, if there are, how much variance each cycle accounts for. See Figure 9.10.

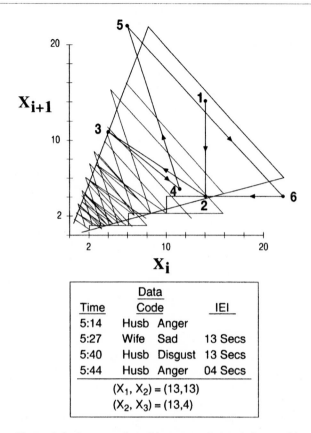

	Data	
Time	Code	IEI
5:14	Husb Anger	
5:27	Wife Sad	13 Secs
5:40	Husb Disgust	13 Secs
5:44	Husb Anger	04 Secs
	$(X_1, X_2) = (13,13)$	
	$(X_2, X_3) = (13,4)$	

Figure 9.8. A scatterplot of interevent interval times with consecutive points connected.

Note that the overall spectral analysis of all the data reveals very little. There seem to be multiple peaks in the data, some representing slower and some faster cycles. However, if we divide the interaction into parts, we can see that there is actually a systematic shift in the cyclicities. The cycle length is 17.5 seconds at first, and then moves to 13.2 seconds, and then to 11.8 seconds. This means that the time for the system to cycle between negative affects is getting shorter as the interaction proceeds. This is exactly what we observed in the state space diagram in which all the points were connected. Hence, in two separate analyses of these data we have been led to the conclusion that this system is not regulated, but is moving toward more and more rapid response times between negative affects. From the data we have available, this interaction seems very negative, relentless, and unabated. Of course, there may be a more macro-level regulation that we

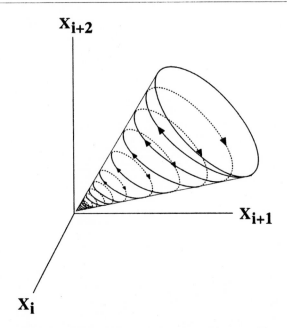

Figure 9.9. System is being drawn toward faster and faster response times in the IEI of negative affect.

do not see that will move the system out toward the base of the cone once it has moved in, and it may oscillate in this fashion. But we cannot know this. At the moment it seems fair to conclude that this unhappy marriage represents a runaway system.

There are lots of other possibilities for what phase-space flow diagrams might look like. One common example is that the data seem to hover quite close to one or two of what are called "steady states." This means that the data really are quite stable, except for minor and fairly random variations. Another common example is that the data seem to move in something like a circle or ellipse around a steady state. The circle pattern suggests one cyclical oscillation. More complex patterns are possible, including chaos (see Gottman, 1990, for a discussion of chaos theory applied to families); we should caution the reader that despite the strange romantic appeal that the chaos theory has enjoyed, chaotic patterns are actually almost never observed. Gottman (1990) suggested that the cyclical phase-space plot was like a steadily oscillating pendulum. If a pendulum is steadily oscillating, like the pendulum of a grandfather clock, energy is constantly being supplied to drive the pendulum, or it would run down (in phase space, it would spiral in toward a fixed point).

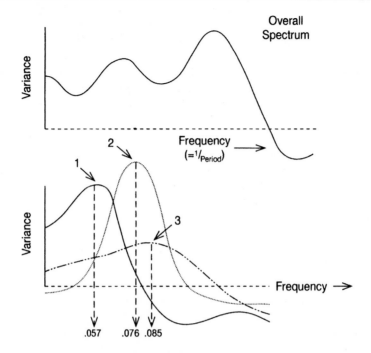

Figure 9.10. Spectrum for three segments showing higher frequencies and shorter periods.

9.6 Summary

When successive events have been coded (or when a record of successive events is extracted from more detailed data), event-sequential data result. When successive intervals have been coded, interval-sequential data result. And when event or state times have been recorded, the result is timed-event or state data. At one level, the representation of these data used by the GSEQ program, these four kinds of data are identical: All consist of successive bins, where bins are defined by the appropriate unit (event, time, or interval) and may contain one, more, or no codes.

In general, all the analytic techniques that apply to event sequences can be applied to state, timed-event, and interval sequences as well, but there are some cautions.

Primarily, the connection, or lack of connection, between coders' decisions and representational units is important to keep in mind and emphasize when interpreting sequential results because in some cases units represent decisions and in other cases arbitrary time units. It is also important to keep in mind how different forms of data representation suit different questions.

When time is an issue – whether percent of time devoted to certain activities or average bout lengths for those activities – time must be recorded and represented, probably using either state or timed-event sequences. When the simple sequencing of events is an issue, event sequences may suffice, or an event-sequential version of state or timed-event sequences created (this can be done with the GSEQ program). Finally, when several dimensions of an event are important, data can be represented using the interval sequential form but, in this case, might better be termed multidimensional event sequential data.

10

Analyzing cross-classified events

10.1 Describing cross-classified events

Earlier, in chapter 3, we distinguished between continuous and intermittent recording strategies, noting that because this is a book about sequential analysis, we would emphasize continuous approaches. Four of these strategies (coding events, recording onset and offset times, timing pattern changes, coding intervals) are continuous both in the sense that observers are continuously alert, ready to record whenever required, and in the sense that the data recorded reflect a continuous record of some aspect of the passing stream of behavior. Such data are typically represented as event, state, timed-event, or interval sequences (see chapter 5). We also mentioned a fifth strategy (cross-classifying events), which is continuous only in the sense that observers are continuously alert, ready to record information about certain events whenever they occur. This strategy results in sequential data only if the information coded represents logically sequential aspects of the event.

This is an appealing strategy for a number of reasons. When an investigator knows beforehand that a particular kind of event is of interest, this approach provides focused information about that event with little extraneous detail. Moreover, the data can be represented and analyzed in simple, well-known, and quite straightforward ways.

As an example, let us again use the Bakeman and Brownlee (1982) study of object struggles, referred to earlier. The event of interest, of course, was an object conflict or, to put it in more neutral terms, a "possession episode." As they defined it, a possession episode required that one child (the taker) touch and attempt to take away an object currently being held by another child (the holder). Observations took place in both a toddler (ages ranged from 12 to 24 months) and a preschool (ages ranged from 40 to 48 months) classroom. In each, a target or focal child strategy was used, that is, observers would focus first on one child for a period of time (in this case, 5 minutes), then on another, switching their attention from child to child according to a predetermined random order. Each time the current

Table 10.1. *Resistance cross-classified by prior possession and dominance for toddlers and preschoolers*

Age group	Dominance	Prior possession	Resistance Yes	No
Toddlers	Yes	Yes	19	7
		No	42	30
	No	Yes	16	4
		No	61	13
Preschoolers	Yes	Yes	6	5
		No	18	5
	No	Yes	9	6
		No	27	4

Note: Counts for the toddlers are based on 20.7 hours of observations; for the preschoolers, on 16.6 hours; see Bakeman and Brownlee (1982).

target child was involved in a possession episode, observers recorded (a) the name of the taker and the name of the holder; (b) whether the taker had had "prior possession," that is, had played with the contested object within the previous minute; (c) whether the holder resisted the attempted take; and (d) whether the taker was successful in gaining the object.

In addition, a linear dominance hierarchy was determined for the children in each class and so, for each episode, it was possible to determine whether the taker was dominant to the holder. Thus each episode was cross-classified as follows: taker dominant (yes/no), taker had prior possession (yes/no), taker resisted (yes/no), and taker successful (yes/no). These were viewed as sequential, in the order given, although, properly speaking, dominance is a "background variable" and not a sequential aspect of the possession episode.

Bakeman and Brownlee chose to regard both resistance and success as outcomes, as "response variables" to be accounted for by the "explanatory variables" of dominance and prior possession. As a result, they presented their data as a series of 2 × 2 × 2 tables (dominance × prior possession × success or dominance × prior possession × resistance) instead of including resistance and success as dimensions in the same table. Because our purpose here is to discuss, in a general way, how to describe (and analyze) cross-classified event data, we shall use an an example just data on dominance, prior possession, and resistance, ignoring success.

Describing cross-classified event data is almost embarrassingly simple. According to time-honored tradition, such data are presented as contingency tables (see Table 10.1). Data can be presented as raw counts or tallies as here, or various percentages can be computed. For example, 72% of the possession episodes met with resistance among toddlers; the corresponding percentage for preschoolers was 75%, almost the same. We could then go on and report percentages of possession episodes encountering resistance when the taker had had prior possession, was dominant, etc., but a clearer way to present such data is as conditional probabilities.

The simple probability that a take attempt would encounter resistance was .72 among toddlers. The conditional probability for resistance, given that the toddler taker had had prior possession, $p(R/P)$, was actually somewhat higher, .76, but the conditional probability for resistance, given that the taker was dominant, $p(R/D)$, dropped to .62. This suggests that, at least among toddlers, the likelihood that the taker will encounter resistance is affected by dominance but not by prior possession.

Among preschoolers, the situation appears reversed. For them, the simple probability that a take attempt would meet with resistance was .75. The conditional probability for resistance, given that the preschooler taker had had prior possession, was only .58, whereas the conditional probability for resistance, given that the taker was dominant was .71, close to the simple or unconditional probability for resistance. Thus preschool takers were less likely to encounter resistance when attempting to "take back" a contested object – as though prior possession conferred some current right. However, this presumed "right" was evident only among the preschoolers, not the toddlers. (The reader may want to verify that the conditional probability values were correctly computed from the data given in Table 10.1.)

This is only descriptive data, however. The question now is, are the differences between simple and conditional probabilities just described larger than one would expect, based on a no-effects or chance model?

10.2 Log-linear models: A simple example

Cross-classified categorical data like those in Table 10.1 can be analyzed relatively easily with what is usually called a log-linear modeling approach (e.g., see Bakeman & Robinson, 1994; Wickens, 1989; for applications see Bakeman, Adamson, & Strisik, 1989, 1995; Bakeman, Forthman, & Perkins, 1992). In principle, this approach is fairly simple and flexible. In addition, results can be expressed in familiar analysis-of-variance-like terminology. Finally, few assumptions need to be made about the data. For all these reasons, this approach, or some variety of it, can prove useful.

Although there are a variety of different kinds of models (e.g., log-linear, logistic, etc.), the general approach is the same. The investigator defines a set of hierarchical modes ("hierarchical" in the sense that simpler models are proper subsets of more complex ones). The simplest model – the null or equiprobale model – contains no terms at all and generates the same expected value for each cell in the contingency table. The most complex model – the saturated model – contains sufficient terms to generate expected values for each cell that are identical to the values actually observed. The idea is to find the least complex model that nonetheless generates expected values not too discrepant from the observed ones, as determined by a goodness-of-fit test. Sometimes the investigator begins with the null or some other simple model. If the data generated by it fail to fit the actual data, then more complex models are tried. If all else fails, the saturated model will always fit the data because it generates values identical to the observed ones. Alternatively, and more typically, the investigator begins with the saturated model and deletes terms until the simplest fitting model is found.

The simplest example of this logic is provided by the familiar chi-square test of independence, although introductory textbooks rarely present it in these terms. The model typically tested first – the no-interaction model – consists of two terms, one for the row variable and one for the column variable. In other words, the row and the column totals for the data generated by the model are forced to agree with the row and column totals actually observed. In introductory statistics, students almost always learn how to compute these expected frequencies, although they are rarely taught to think of them as generated by a particular model.   .

For example, if we look just at the 2 × 2 table detailing prior possession and resistance for dominant toddlers (shown in the upper right-hand corner of Table 10.1), most readers would have no trouble computing the expected frequencies. The total tally for row 1 is 26 (19 + 7), for row 2, 72; similarly, the total for column 1 is 61 (19 + 42), for column 2, 37; the grand total is 98. Thus the expected frequencies are 16.2 (61 times 26 divided by 98) and 9.8 for row 1 and 44.8 and 27.2 for row 2. In this case, the generated data fit the observed quite well, and the chi-square would be small. In the context of a chi-square analysis just of this 2 × 2 table, the investigator would conclude that, among dominant toddlers, prior possession and resistance are not related.

However, if chi-square were big, meaning that data did not fit this particular model, the investigator would need to invoke a more complex model. In the case of a 2 × 2 table, the only model more complex than the no-interaction model is the saturated model, which includes (in addition to the row and column effects) an interaction term. To recapitulate, the expected frequencies computed for the common chi-square test of independence are in fact generated by a no-interaction model. If the chi-square is significantly large, then this model fails to fit the data. An interaction term is required, which means that the row and column variables are not independent of each other, but are in fact associated. Often this lack of independence is exactly what the investigator hopes to demonstrate.

Following usual notational conventions (e.g., Fienberg, 1980), the no-interaction model mentioned in the previous paragraph would be indicated as the [R] [C] model, meaning that cell values are constrained by the model to reflect just the row and column totals. The saturated model would be represented as [R] [C] [RC], meaning that in addition to the row and column constraints, the cell totals must also reflect the row × column [RC] cross-classification totals. In the case of a two-dimensional table, this means that the model is saturated and in fact generates data identical with the observed. Because meeting the [RC] constraints means that necessarily the [R] and [C] constraints are met, the saturated model is usually indicated simply by [RC]. In general, to simplify the notation, "lower"-level terms implied by higher-order terms are usually not included when particular models are described. They are simply assumed.

At this point, we hope the reader has a clear idea of the logic of this approach, at least with respect to two-dimensional contingency tables. The situation with respect to tables with three or more dimensions is somewhat more complex. For one thing, computing expected frequencies for various models is not the simple matter it is with two-dimensional tables; usually, computer programs are used. For another, the notation can become a little confusing. Our plan here is to demonstrate the approach with the three-dimensional dominance × prior possession × resistance tables for toddlers and preschoolers, as given in Table 10.1. This hardly exhausts the topic, but we hope it will given interested readers a start, and that they will then move on to more detailed and specialized literature (e.g., Bakeman & Robinson, 1994).

For the 2 × 2 × 2 tables for the toddlers and preschoolers, it seems clear that the null or equiprobable model (all cells have the same value) would not fit these data. In fact, we would probably not test the null model first, but, as with the chi-square test of independence, would begin with a more complex one. Earlier we said that resistance (R) was regarded as a response or outcome variable, and that dominance (D) and prior possession (P) were regarded as explanatory variables. That is, substantively we want to find out if dominance and/or prior possession affected resistance. Thus we would begin testing with the [R] [DP] model. This model simply constrains the generated data to reflect the resistance rates and the dominance cross-classified by prior possession rates actually observed. In particular, it does not contain any terms that suggest that either dominance, or prior possession, or their interaction, is related to the amount of resistance encountered.

In analysis-of-variance terms, the [R][DP] model is the "no effects" model. In a sense, the [DP] term just states the design, whereas the fact that the response variable, [R], is not combined with any of the explanatory variables indicates that none affect it. If the [R][DP] model failed to fit the data, but the [RD][DP] model did, we would conclude that there was a main effect for dominance – that unless dominance is taken into account, we fail to make very good predictions for how often resistance will occur. Similarly, if the [RP][DP] model fit the data, we would conclude that there was a main effect for prior possession. If the [RD][RP][DP] model fit, main effects for both dominance and prior possession would be indicated. Finally, if only the [RDP] model fit the data (the saturated model), we would conclude that, in order to account for resistance, the interaction between dominance and prior possession must be taken into account.

In the present case, the no-effects model failed to fit the observed data. For both toddlers and preschoolers, the chi-square comparing generated to observed was large and significant (values were 11.3 and 7.2, $df = 3$, for toddlers and preschoolers, respectively; these are likelihood-ratio-chi-squares, computed by Bakeman & Robinson's [1994] ILOG program). However, for toddlers the [RD][DP] model, and for preschoolers the [RP][DP] model generated data quite similar to the observed (chi-squares were 1.9 and 0.8, $df = 2$, for toddlers and preschoolers, respectively; these chi-squares are both insignificant, although in both cases the difference between them and the no-effects model is significant). This is analogous to a main effect for dominance among toddlers and a main effect for prior possession among preschoolers. In other words, the dominance of the taker affected whether his or her take attempt would be resisted among toddlers, but whether the taker had had prior possession of the contested object or not affected whether he or she would meet with resistance among preschoolers. Thus the effects noted descriptively in the previous section are indeed statistically significant. Bakeman and Brownlee (1982) interpreted this as evidence for shared possession rules, rules that emerge somewhere between 2 and 3 years of age.

10.3 Summary

Sometimes investigators who collect observational data and who are interested in sequential elements of the behavior observed seem compelled both to obtain a continuous record of selected aspects of the passing stream of behavior and to analyze exhaustively that continuous record. An alternative is to record just selected aspects of certain kinds of events. The kind of event is defined beforehand (for example, possession struggles) as well as the aspects of interest. Each aspect corresponds to a codable dimension. In the case of the possession struggles described in the previous sections, the dimensions were dominance, prior possession, and resistance. The codes for each dimension were the same (either yes or no), but in other cases the mutually exclusive and exhaustive codes for the various dimensions could well be different. In all cases, cross-classified event data result.

A major advantage of such data is, first, cross-classified categorical data are hardly exotic or new, and second, ways of analyzing such data are relatively well understood (e.g., Bakeman & Robinson, 1994; Kennedy, 1983, 1992; Upton, 1978; Wickens, 1989). In this chapter, a simple example was given, showing how log-linear results can be expressed in analysis-of-variance-like terms. The technique, however, is not confined just to contingency tables, but can also be applied to event-sequence and time-sequence data; this was mentioned earlier in section 7.6. The interested reader is advised to consult Bakeman and Quera (1995b) and Castellan (1979).

11

Epilogue

11.1 Kepler and Brahe

When Johannes Kepler went to work for the astronomer Tycho Brahe, he found himself in the midst of a strange set of circumstances. Brahe lived in a castle on an island, and there he conducted his painstakingly careful observations. Brahe had a dwarf who scrambled for food in the dining hall. Brahe also had a silver nose that replaced the natural nose he had lost in a duel. Kepler was given the problem of computing the orbit of Mars, and he wrote in one of his letters that Brahe fed him data in little bits, just as the dwarf received crumbs under the table.

Kepler had come to Brahe because of the observations. He was hungry for the observations. Without them, the principles of planetary motion – the patterns he discovered – would never have been discovered. Without these patterns, Newton's explanation – the universal theory of gravitation – would never have emerged.

We need to observe, and we need to do it very well, with imagination, with boldness, and with dedication. If we do not observe, we shall never see what is there. If we never see what is there, we shall never see the patterns in what is there. Without the patterns, there will never be the kind of theory that we can build with.

Observing and discovering pattern is what this book is about. We do this kind of thing for a living, and we have chosen to do it because it is what we think science is about. Obviously we think it is not really that hard to do. But we are lonely. We want company in this enterprise. Only about 8% of all psychological research is based on any kind of observation. A fraction of that is programmatic research. And, a fraction of that is sequential in its thinking.

This will not do. Those of us who are applying these new methods of observational research are having great success. We are beginning to find consistent results in areas previously recalcitrant to quantitative analysis: how babies learn to interact with their parents and organize their behavior; how young children make friends or are rejected by their peers; how marriages succeed or fail by how spouses interact; how families weather stress or create pathology.

In many senses, in developmental, social, and clinical psychology we are returning to earlier efforts with a new conceptual technology, and it is paying off. We are able to return to the spirit of observation that characterized developmental psychology in the 1930s and ask questions about social and emotional development more clearly. We are now able to return to initial clinical theorizing about families in the 1950s and bring a great deal of precision to the task.

What is new is not a new statistical technique. It is a whole new way of thinking about interaction in terms of its temporal form, about pattern that occurs and recurs in time. We believe that great conceptual clarity can be obtained by thinking about temporal patterning, and we believe that anyone who has collected observational data over time and ignores time is missing an opportunity.

Kepler needed the observations. The observations needed Kepler.

11.2 Soskin and John on marital interaction

Over 30 years ago, Roger Barker edited a classic book titled *The Stream of Behavior*, which was seminal in the history of naturalistic observational research. The book also focused on social interaction. We believe that it would be useful to review a chapter in that book by Soskin and John that tackled the description of marital interaction. We shall compare their analysis with what we currently know about marital interaction. Our objective is to make the point that we can now confront old problems with new tools.

Soskin and John (1963) spent a year pilot-testing a 3-pound radio transmitter to be worn in a kind of backpack arrangement by two young husband–wife pairs. The subjects received an expense-free vacation in a resort community; they lived in a small cottage at the edge of a large lake, though they mingled freely with other guests. The transmitters could be disconnected by the subjects if they felt the need for privacy, but neither couple did so during the day. Transmitters were turned off after midnight. Soskin and John presented an extensive fragment of one couple's conversation while they were rowing. The episode contains good natural humor:

> Jock: Yo-ho, heave ho. You do the rowing.
> Roz: Nuts to that idea. You're a big strong man. Mmmm!
> Jock: Yeah, but I have a handicap.
> Roz: Yeah, you have a handicap in your head.

The episode also contains come conflict:

> Roz: You're dipping your oars too deep, dear.
> Jock: I can't feather them, either.

Roz: You should be able to . . .
Jock: Change places with me.
Roz: Let me get this off first.
Jock: No, change places first. Hold it up and . . . you can stand up. It's perfectly all right.
Roz: That's the first thing I learned about water safety, love. Don't stand up in a boat . . .
Jock: Well I wanted you to stay into the waves, whatever you did.
Roz: Well, why didn't you tell me so!
Jock: Go up that way.
Roz: (slightly irritated) Which way *do* you want to go?
Jock: This way.

Soskin and John analyzed their corpus of data in three ways: (a) a "structural analysis"; (b) a "functional analysis"; and, (c) a "dynamic analysis." The structural analysis primarily computed talk times. The functional analysis began by distinguishing between "informational" messages ("It's four miles from here") and "relational" messages ("Then get going"). They wrote that relational talk "encompasses the range of verbal acts by which a speaker manages his interpersonal relations" (p. 253).

In their functional analysis they eventually identified six categories of messages: (a) expressive statements ("Ouch!" "Wow!" "Darn!"); (b) "excogitative" statements, which are "most commonly described as 'thinking aloud' "(p. 255) ("Humm . . . what have I done wrong here?" "Oh, I see!"); (c) "signomes," which are "messages that report the speaker's present physical or psychological state" (p. 255) ("I'm cold!" "Oh, Jock, I like that!"); (d) "metrones," which are evaluative statements arising out of the subject's belief system ("What a fool I've been!" "You shouldn't do that!"); (e) "regones," which control or channel the behavior of the listener ("Why don't you do it right now?"); and (f) "structones," which include information statements ("I weigh 181 pounds").

The dynamic analysis distinguished three variables: state, locus-direction, and bond. State involved affective information: (a) joy, glee, high pleasure; (b) satisfaction, contentment, liking; (c) ambivalence; (d) mild apprehension, dislike, frustration, disappointment; (e) pain, anger, fear, grief; and (f) neutrality. The locus-direction variable was indicated with arrows up or down "for the direction and locus of state change it would produce from the point of view of a neutral observer" (p. 258). These categories were (a) wants, wishes, self-praise; (b) mutually complimentary statements; (c) derogation, reproof, rebuffs, which imply the speaker's superiority; (d) self-criticism; (e) apology, praise; (f) compliments, permission; (g) mutually unfavorable statements; (h) accusations, reproof; and (i) no inferable change. "Bonds" referred to "the degree of intimacy the speaker was willing to tolerate in the relationship" (p. 258).

Soskin and John reported several findings based on their analysis of their tapes. The structural analysis showed that, in most situations, Jock talked a lot and could be described as highly gregarious (he talked about 36% of the total talk time in a four-person group). His longer utterances were predominantly structones (factual-information exchanges).

The functional analysis of 1850 messages of Roz and Jock's talk showed that Roz was significantly more expressive (8.6% vs. Jock's 3.8%), less controlling (fewer regones: 11.0% vs. Jock's 13.9%), and less informational (fewer structones: 24.5% vs. Jock's 31.3%). They concluded:

> Roz produced a high percentage of expressive messages whenever the two were out of the public view and became noticeably more controlled in the presence of others. Jock's output, on the other hand, was relatively low throughout. (p. 267)

This is not quite consistent with the earlier analysis of Jock as gregarious and a high-output talker. They then turned to the results of their dynamic analysis, which they began describing as follows:

> The very dimensions by which it was hoped to identify inter- and intrapersonal changes in the sequential development of an episode proved most difficult to isolate. (p 267)

Unfortunately, they could find no consistent patterns in the way Roz and Jock tried to influence and control one another. They wrote:

> The very subtle shifts and variations in the way in which these two people attempted to modify each other's states sequentially throughout this episode obliged us to question whether summaries of very long segments of a record reflect the actual sequential dynamics of the behavior in a given episode. (p. 268)

Equally disappointing was their analysis of affect; they wrote:

> As with locus-direction shift, the assessment of affective state changes met with only marginal success. (p. 270)

However, they did conclude that:

> the coders saw Roz as producing a higher percent of mildly negative statements than her husband in all five of the episodes, in three of which the difference was statistically significant. By contrast, in all five episodes Jock was seen as producing a higher percent of neutral statements, and in four of the five episodes the difference between them was significant. (p. 272)

In these results, we can see a struggle with coding systems that are unwieldy, that are hard to fit together, and that lack a clear purpose or focus. The research questions are missing.

11.3 Marital interaction research since 1963

In the early 1970s, psychologists began systematically applying observational methods to the study of marital interaction. The first important such study was Raush, Barry, Hertel, and Swain's (1974) book *Communication, Conflict, and Marriage*. This was a longitudinal study that followed couples from a newlywed stage through the pregnancy and birth of the couple's first child. There were major research questions: (1) Was there consistency over time in couples' interactional style in resolving conflict? (2) How did the pregnancy period affect the marriage? (3) What sex differences exist? (4) How are happily and unhappily married couples different in the way they resolve conflict? Raush et al. employed one coding system, a series of improvised conflict situations, and they used sequential analysis (employing multivariate information theory).

Raush et al. found that couples had an interactive style that was consistent over time. In a later study, Gottman (1980b) found that there was greater cross-situational consistency within couples (from high- to low-conflict interactions) when sequential z scores rather than unconditional probabilities were used.

Contrary to what Soskin and John reported about Jock and Roz, Raush et al. found no evidence to support the contention that in marriages men were less expressive and more instrumental than women. However, consistent with Sokin and John's affect results, they did report that women were more coercive than men, whereas men were more reconciling. These sex differences were enhanced during the pregnancy phase. However, in general, "discordant" couples were far more coercive than "harmonious" couples.

Subsequent sequential analytic research on the question of differences between satisfied and dissatisfied couples has produced a clear picture in conflict-resolution strategies. There is now a body of literature that can be called upon. We shall not review this literature here because our purposes are methodological. For a recent review, see Noller (1984). One series of studies with one coding system was reported in a research monograph by Gottman (1979a). We shall summarize the consistent results of these studies by giving examples of conversation sequences.

Historically, three methodological innovations were necessary before this work could proceed: appropriate observational methods, sequential analysis, and the systematic study of affect displayed through nonverbal behavior. Consider the following dialogue of a dissatisfied couple discussing how their day went (H = husband, W = wife; extracts are from unpublished transcripts):

H: You'll never guess who I saw today, Frank Dugan.
W: So, big deal, you saw Frank Dugan.

H: Don't you remember I had that argument with him last week?
W: I forgot.
H: Yeah.
W: So I'm sorry I forgot, all right?
H: So it is a big deal to him.
W: So what do you want me to do, jump up and down?
H: Well, how was your day, honey?
W: Oh brother, here we go again.
H: (pause) You don't have to look at me that way.
W: So what d'ya want me to do, put a paper bag over my head?

Using the Couples Interaction Scoring System (CISS), which codes both verbal and nonverbal behaviors of both speaker and listener, Gottman and his associates coded the interaction of couples who were satisfied or dissatisfied with their marriages. Among other tasks, couples were studied attempting to resolve a major area of disagreement in their marriages.

Gottman's results

The major question in this research was "what were the differences between satisfied and dissatisfied couples in the way they resolve conflict?"

Basically, these differences can be described by using the analogy of a chess game. A chess game has three phases: the beginning game, the middle game, and the end game. Each phase has characteristic good and bad maneuvers and objectives. The objectives can, in fact, be derived inductively from the maneuvers. The goal of the beginning phase is control of the center of the chessboard and development of position. The goal of the middle game is the favorable exchange of pieces. The goal of the end game is checkmate. Similarly, there are three phases in the discussion of a marital issue. The first phase is "agenda-building," the objective of which is to get the issues out as they are viewed by each partner. The second phase is the "arguing phase," the goal of which is for partners to argue energetically for their points of view and for each partner to understand the areas of disagreement between them. The third phase is the "negotiation," the goal of which is compromise.

It is possible to discriminate the interaction of satisfied and dissatisfied couples in each phase. In the agenda-building phase, cross-complaining sequences characterize dissatisfied couples. A cross-complaining sequence is one in which a complaint by one person is followed by a countercomplaint by the other. For example:

W: I'm tired of spending all my time on the housework. You're not doing your share.
H: If you used your time efficiently you wouldn't be tired.

A validation sequence recognizes the potential validity of the other person's viewpoint before making a counterstatement. Usually the validation sequence differs from the cross-complaining sequence by the use of "assent codes" such as "Yeah," "Oh," "Mmmhmmm," and so on. For example, below is a cross-complaining sequence followed by the same exchange as a validation sequence.

Cross-complaining:

> W: I've been home alone all day, cooped up with the kids.
> H: I come home tired and just want to relax.

Validation:

> W: I've been home alone all day.
> H: Uh-huh.
> W: Cooped up with the kids.
> H: Yeah, I come home tired.
> W: Mmm.
> H: And just want to relax.
> W: Yeah.

In the negotiation phase, counterproposal sequences characterize the interaction of dissatisfied couples, whereas contracting sequences characterize the interaction of satisfied couples. In a counterproposal sequence, a proposal by one partner is met immediately by a proposal by the other partner, whereas in a contracting sequence there is first some acceptance of the partner's proposal. The agreement codes that discriminate counterproposal and contracting sequences are very different from those that discriminate cross-complaining and validation sequences. Instead of simple agreement or assent, contracting sequences include direct modification of one's own view:

Counterproposal:

> W: We spent all of Christmas at your mother's last year. This time let's spend Christmas at my mother's.
> H: Let's spend it again at my mother's this year. It's too late to change it. We can discuss our plans for next year now.

Contracting:

> W: We spent all of Christmas at your mother's last year. This time let's spend Christmas at my mother's.
> H: Yeah you're right, that's not fair. How about 50-50 this year?

At some points the conversation of the two groups of couples would be indistinguishable without the use of nonverbal codes.

The anatomy of negative affect

The deescalation of negative affect, not the reciprocation of positive affect (known in the literature on marital therapy as the quid pro quo hypothesis), discriminated happy from unhappy marriages in these studies. Another finding concerned the role of statements about the process of communication (metacommunication), such as "You're interrupting me." There were no differences in the amount of metacommunication between satisfied and dissatisfied couples, but the sequences in the two groups differed markedly. Metacommunication tends to be what is called, in Markov model theory, an "absorbing state" for unhappily married couples, i.e., it becomes nearly impossible to exit once entered. For satisfied couples, metacommunicative chains are brief and contain agreements that lead rapidly to other codes. For example, a metacommunicative chain in a satisfied marriage might be:

> H: You're interrupting me.
> W: Sorry, what were you saying?
> H: I was saying we should take separate vacations this year.

For a dissatisfied couple the chain might be:

> H: You're interrupting me.
> W: I wouldn't have to if I could get a word in edgewise.
> H: Oh, now I talk too much. Maybe you'd like me never to say anything.
> W: Be nice for a change.
> H: Then you'd never have to listen to me, which you never do anyway.
> W: If you'd say something instead of jabbering all the time maybe I would listen.

It is not the amount of metacommunication, but how it is delivered that determines the sequence that follows and whether its role facilitates communication. This fact could not have been discovered without a sequential analysis of the data. Note that what makes the metacommunication effective is that it changes the affective nature of the interaction. If the affect simply transfers to the metacommunication, it cannot function as a repair mechanism.

Another pattern common to both satisfied and dissatisfied couples is called "mind reading" – making attributions of emotions, opinions, states of mind, etc., to a spouse. The effect of mind reading depends entirely on the affect with which it is delivered. If mind reading is delivered with neutral or positive affect, it is responded to as if it were a question about feelings; it is agreed with and elaborated upon, usually with neutral affect:

> H: You always get tense at my mother's house.
> W: Yes, I do. I think she does a lot to make me tense.

If mind reading is delivered with negative affect, it is responded to as if it were a criticism; it is disagreed with and elaborated upon, usually with negative affect:

> H: You always get tense at my mother's house.
> W: I do not. I start off relaxed until she starts criticizing me and you take her side.

Satisfied couples continually intersperse various subcodes of agreement into their sequences. In the agenda-building phase, this is primarily a simple "assent" form of agreement, as in "oh yeah," "uh huh," "I see," and so on, whereas in the negotiation phase, this is primarily actual acceptance of the other's point of view and modification of one's own point of view. These listener responses or "backchanneling" (Duncan & Fiske, 1977) are clear communications to the speaker that the listener is "tracking" the speaker. But they do more than regulate turns, especially in the beginning phases of marital conflict resolution. They communicate agreement not with the speaker's point of view or content, but with the speaker's affect. By indicating that it might make sense to see things as the other does, these responses grease the wheels for affective expression.

In the negotiation phase of the discussion, the agreement codes are very different. They are not "assent," but direct agreement with the other's point of view ("Yes, you're right," or "I agree with that"). They may even involve accepting some modification of one's own point of view in order to reach a solution to the problem. This creates a climate of agreement that has profound consequences for the quality of the interaction.

We see in all these results that in dissatisfied marriages couples are far more defensive and less receptive to their partners. To investigate the nature of this defensiveness, Robert Levenson and John Gottman began collecting autonomic nervous system (ANS) data during marital interaction. They discovered that ANS arousal during conflict resolution is *highly* predictive (simple correlations in the 90s) of changes in marital satisfaction over a 3-year longitudinal period, controlling initial levels of marital satisfaction. Here we have the possibility of a theoretical basis for the observational results. What are the setting conditions of ANS arousal? Will all discussions of disagreements produce arousal? What are the consequences of ANS arousal? Are there sex differences in ANS arousal in response to intense negative affect? These questions are currently being pursued in Gottman's and Levenson's laboratories.

To summarize, in this section we have suggested that we are now in a position to tackle old and venerable questions with new tools. Observational techniques have been successful in the study of marital interaction (and in other areas) in *identifying stable phenomena*. This is the first step toward the construction of theory, which seeks elegant, far-reaching, and parsimonious *explanations* for the observed phenomena.

Just as Kepler needed Brahe's observations, the observations needed Kepler. Kepler found the patterns. Newton explained them with the theory of gravitation. We need our Brahes, our Keplers, and our Newtons. This book is an advertisement for all of the "instrument makers" who have developed observational methodology to its current handy state.

Appendix: A Pascal program to compute kappa and weighted kappa

This appendix contains the source code for a Pascal program that computes kappa and weighted kappa. The program was compiled using Borland's Pascal 6.0 and should run on IBM-compatible microcomputers in a Windows or DOS environment. The program contains essentially no error checking for the numbers the user enters, so these must be correct. All values entered (the number of rows, weights if any, the tallies themselves) are integers, so entering a letter or even a decimal point instead of a digit, for example, will cause the program to fail (unless error checking code is added).

```
Program Kappa; { A simple Pascal no-bells, no-whistles,}
                { no-errors-permitted-in-input program  }
uses  Crt;      { to compute kappa and weighted kappa. }
                { (c) Roger Bakeman, September 1995.    }
const Max = 20;
      Enter = chr(13);
      N : word = 0;

var   i, j : word;
      Ch    : char;
      X, W : array [1..Max,1..Max] of word;
      WeightsDefined : boolean;

{- - - - - - - - - - - - - - - - - - - - - - - - - - - - - }
Procedure ComputeKappa;
var   i, j : word;
      M : array[1..Max+1,1..Max+1] of real;
      Kappa, Numer, Denom : real;

begin
  for i := 1 to N do     { Set row & col sums to zero. }
  begin
    M[i,N+1] := 0.0;
```

```
   M[N+1,i]  := 0.0;
   end;
   M[N+1,N+1] := 0.0;
                          { Tally row & col totals.        }
   for i := 1 to N do for j := 1 to N do
   begin
     M[i,N+1]  := M[i,N+1] + X[i,j];
     M[N+1,j]  := M[N+1,j] + X[i,j];
     M[N+1,N+1] := M[N+1,N+1] + X[i,j];
   end;
                            { Compute exp. frequencies.    }
   for i := 1 to N do for j := 1 to N do
     M[i,j] := M[i,N+1] * M[N+1,j] / M[N+1,N+1];

   Numer := 0.0;           { Compute kappa.                }
   Denom := 0.0;
   for i := 1 to N do for j := 1 to N do
   begin
     Numer := Numer + W[i,j] * X[i,j];
     Denom := Denom + W[i,j] * M[i,j];
   end;
   Kappa := 1.0 - Numer/Denom;
   writeln ('  Kappa =',Kappa:7:4);
   writeln;
end;


{- - - - - - - - - - - - - - - - - - - - - - - - - - - - - }
Procedure DefaultWeights; { Provide standard weights,  }
var i, j : word;          { 0 on diagonal, 1 otherwise.}

begin
  for  i := 1 to N do for j := 1 to N do
  if (i = j) then W[i,j] := 0 else W[i,j] := 1;
end;

{- - - - - - - - - - - - - - - - - - - - - - - - - - - - - }
Procedure AskForNumbers (Kind : word);
const Message:                  {   Kind=1  Kind=2        }
    array [1..2] of string[7] = ('weights', 'tallies');
var i : word;

begin
```

```
    writeln ('  Enter  ',N:3,'   ',Message[Kind],
            '  (with spaces between for) ...');
    for i :=1 to N do
    begin                              {Read numbers;       }
      write (' row',i:3,'> ');         {weights if Kind=1,}
      for j := 1 to N do read (X[i,j]);{tallies if Kind=2.}
      readln;                          {Must be integers. }
    end;
    if (Kind = 1) then WeightsDefined := true;
  end;

{- - - - - - - - - - - - - - - - - - - - - - - - - - - - - }
Function WantsSame (Kind : word) : boolean;
const Message : array [1..5] of string[12] =
      ('Same # rows ',
       'Same labels ',    { Ask Y|N questions. Return  }
       'Weighted k  ',    { TRUE if Y, y, or Enter.    }
       'Same weights ',
       'More kappas ');
var Ch : char;

begin
  write ('    ',Message[Kind],' (Y|N)? ');
  Ch := ReadKey; writeln (Ch);
  WantsSame := (UpCase(Ch) = 'Y') or (Ch = Enter);
end;

{- - - - - - - - - - - - - - - - - - - - - - - - - - - - - }
Procedure AskForOrder;     { Ask for order of matrix.  }
begin                      { Must be 1 through 20.     }
  repeat
    write ('  Number of rows (1-20)? ');
    readln (N)
  until (N > 0) and (N  <=  20);
  weightsDefined := false;
end;

{- - - - - - - - - - - - - - - - - - - - - - - - - - - - - }
BEGIN
  TextColor (Black) ;
  TextBackground (LightGray) ;
  ClrScr;
```

```
Writeln
('Compute kappa or wt kappa; (c) Roger Bakeman, GSU');

repeat
  if N = 0 then AskForOrder
  else if not WantsSame(1) then AskForOrder;

  if not WeightsDefined
  then begin if WantsSame(3) then AskForNumbers(1) end
  else if not WantsSame(4) then AskForNumbers(1);

  if not WeightsDefined then DefaultWeights;
  AskForNumbers(2) ;
  ComputeKappa;
until not WantsSame(5);
END.
```

References

Adamson, L. B., & Bakeman, R. (1985). Affect and attention: Infants observed with mothers and peers. *Child Development, 56,* 582–593.

Ainsworth, M. D. S., Blehar, M. C., Waters, E., & Wall, S. (1978). *Patterns of attachment.* Hillsdale, NJ: Lawrence Erlbaum.

Allison, P. D., & Liker, J. K. (1982). Analyzing sequential categorical data on dyadic interaction: A comment on Gottman. *Psychological Bulletin, 91,* 393–403.

Anderson, T. W., & Goodman, L. A. (1957). Statistical inference about Markov chains. *Annals of Mathematical Statistics, 28,* 89–110.

Altmann, J. (1974). Observational study of behaviour: Sampling methods. *Behaviour, 49,* 227–267.

Altmann, S. A. (1965). Sociobiology of rhesus monkeys. II. Stochastics of social communication. *Journal of Theoretical Biology, 8,* 490–522.

Attneave, F. (1959). *Applications of information theory to psychology.* New York: Henry Holt.

Bakeman, R. (1978). Untangling streams of behavior: Sequential analysis of observation data. In G. P. Sackett (Ed.), *Observing behavior* (Vol. 2): *Data collection and analysis methods* (pp. 63–78). Baltimore: University Park Press.

Bakeman, R. (1983). Computing lag sequential statistics: The ELAG program. *Behavior Research Methods & Instrumentation, 15,* 530–535.

Bakeman, R. (1992). *Understanding social science statistics: A spreadsheet approach.* Hillsdale, NJ: Erlbaum.

Bakeman, R., & Adamson, L. B. (1984). Coordinating attention to people and objects in mother–infant and peer–infant interaction. *Child Development, 55,* 1278–1289.

Bakeman, R., Adamson, L. B., & Strisik, P. (1989). Lags and logs: Statistical approaches to interaction. In M. H. Bornstein & J. Bruner (Eds.), *Interaction in human development* (pp. 241–260). Hillsdale, NJ: Erlbaum.

Bakeman, R., Adamson, L. B., & Strisik, P. (1995). Lags and logs: Statistical approaches to interaction (SPSS Version). In J. M. Gottman (Ed.), *The analysis of change* (pp. 279–308). Hillsdale, NJ: Erlbaum.

Bakeman, R., & Brown, J. V. (1977). Behavioral dialogues: An approach to the assessment of mother–infant interaction. *Child Development, 49,* 195–203.

Bakeman, R., & Brownlee, J. R. (1980). The strategic use of parallel play: A sequential analysis. *Child Development, 51,* 873–878.

Bakeman, R., & Brownlee, J. R. (1982). Social rules governing object conflicts in toddlers and preschoolers. In K. H. Rubin & H. S. Ross (Eds.), *Peer relationships and social skills in childhood* (pp. 99–111). New York: Springer-Verlag.

Bakeman, R., & Casey, R. L. (1995). Analyzing family interaction: Taking time into account. *Journal of Family Psychology, 9,* 131–143.

Bakeman, R., & Dabbs, J. M., Jr. (1976). Social interaction observed: Some approaches to the analysis of behavior streams. *Personality and Social Psychology Bulletin, 2,* 335–345.

Bakeman, R., & Dorval, B. (1989). The distinction between sampling independence and empirical independence in sequential analysis. *Behavioral Assessment, 11,* 31–37.

Bakeman, R., Forthman, D. L., & Perkins, L. A. (1992). Time-budget data: Log-linear and analysis of variance compared. *Zoo Biology, 11,* 271–284.

Bakeman, R., McArthur, D., & Quera, V. (1996). Detecting group differences in sequential association using sampled permutations: Log odds, kappa, and phi compared. *Behavior Research Methods, Instruments, and Computers, 28,* 446–457.

Bakeman, R., & Quera, V. (1992). SDIS: A sequential data interchange standard. *Behavior Research Methods, Instruments, and Computers, 24,* 554–559.

Bakeman, R., & Quera, V. (1995a). *Analyzing interaction: Sequential analysis with SDIS and GSEQ.* New York: Cambridge University Press.

Bakeman, R., & Quera, V. (1995b). Log-linear approaches to lag-sequential analysis when consecutive codes may and cannot repeat. *Psychological Bulletin, 118,* 272–284.

Bakeman, R., & Robinson, B. R. (1994). *Understanding log-linear analysis with ILOG: An interactive approach.* Hillsdale, NJ: Erlbaum.

Bakeman, R., Robinson, B. F., & Quera, V. (1996). Testing sequential association: Estimating exact *p* values using sampled permutations. *Psychological Methods. 1,* 4–15.

Bishop, Y. M. M., Fienberg, S. E., & Holland, P. W. (1975). *Discrete multivariate analysis.* Cambridge, MA: MIT Press.

Box, G. E. P., & Jenkins, G. M. (1970). *Time-series analysis: Forecasting and control.* San Francisco: Holden-Day.

Brazelton, T. B., Koslowski, B., & Main, M. (1974). The origins of reciprocity: The early mother–infant interaction. In M. Lewis & L. A. Rosenblum (Eds.), *The effect of the infant on its caregiver* (pp. 49–76). New York: Wiley.

Brennan, R. L. (1983). *Elements of generalizability theory.* Iowa City: ACT Publications.

Brown, J. V., & Bakeman, R. (1980). Relationships of human mothers with their infants during the first year of life: Effects of prematurity. In R. W. Bell & J. P. Smotherman (Eds.), *Maternal influences and early behavior* (pp. 353–373). New York: Spectrum.

Brown, J. V., Bakeman, R., Snyder, P. A. Fredrickson, W. T., Morgan, S. T., & Helper, R. (1975). Interactions of black inner-city mothers with their newborn infants, *Child Development, 46,* 677–686.

Brownlee, J. R., & Bakeman, R. (1981). Hitting in toddler–peer interaction. *Child Development, 52,* 1076–1079.

Cairns, R. B., & Green, J. A. (1979). How to assess personality and social patterns: Observations or ratings? In R. B. Cairns (Ed.), *The analysis of social interactions: Methods, issues, and illustrations* (pp. 209–226). Hillsdale, NJ: Lawrence Erlbaum.

Castellan, N. J. (1979). The analysis of behavior sequences. In R. B. Cairns (Ed.), *The analysis of social interactions: Methods, issues, and illustrations* (pp. 81–116). Hillsdale, NJ: Lawrence Erlbaum.

Chatfield, C. (1973). Statistical inference regarding Markov chain models. *Applied Statistics, 22,* 7–20.

Cohen, J. A. (1960). Coefficient of agreement for nominal scales. *Educational and Psychological Measurement, 20,* 37–46.

Cohen, J. (1968). Weighted kappa: Nominal scale agreement with provision for scaled disagreement or partial credit. *Psychological Bulletin, 70,* 213–220.

Cohen, J. (1977). *Statistical power analysis for the behavioral sciences.* New York: Academic Press.

Cohen, J., & Cohen, P. (1983). *Applied multiple regression/correlation analysis for the behavioral sciences.* Hillsdale, NJ: Lawrence Erlbaum.

Cohn, J. F., & Tronick, E. Z. (1987). Mother–infant face-to-face interaction: The sequence of dyadic states at 3, 6, and 9 months. *Developmental Psychology, 23,* 68–77.

Condon, W. S., & Ogston, W. D. (1967). A segmentation of behavior. *Journal of Psychiatric Research, 5,* 221–235.

Conger, A. J., & Ward, D. G. (1984). Agreement among 2 × 2 agreement indices. *Educational and Psychological Measurement, 44,* 301–314.

Cook, J., Tyson, R., White, J., Rushe, R., Gottman, J., & Murray, J. (1995). The mathematics of marital conflict: Qualitative dynamic modeling of marital interaction. *Journal of Family Psychology, 9,* 110–130.

Cronbach, L. J., Gleser, G. C., Nanda, H., & Rajaratnam, N. (1972). *The dependability of behavioral measurements: Theory of generalizability for scores and profiles.* New York: Wiley.

Crosbie, J. (1995). Interrupted time-series analysis with short series: Why it is problematic; How it can be improved. In J. Gottman (Ed.), *The analysis of change* (pp. 361–398). Hillsdale, NJ: Erlbaum.

Dabbs, J. M., Jr., & Swiedler, T. C. (1983). Group AVTA: A microcomputer system for group voice chronography. *Behavior Research Methods & Instrumentation, 15,* 79–84.

Duncan, S., Jr., & Fiske, D. W. (1977). *Face to face interaction.* Hillsdale, NJ: Lawrence Erlbaum.

Edgington, E. S. (1987). *Randomization tests* (2nd Ed.). New York: Marcel Dekker.

Ekman, P. W., & Friesen, W. (1978). *Manual for the facial action coding system.* Palo Alto, CA: Consulting Psychologist Press.

Fienberg, S. E. (1980). *The analysis of cross-classified categorical data.* Cambridge, MA: MIT Press.

Fleiss, J. L. (1981). *Statistical methods for rates and proportions.* New York: Wiley.

Fleiss, J. L. (1986). *The design and analysis of clinical experiments.* New York: John Wiley.

Fleiss, J. L., Cohen, J., & Everitt, B. S. (1969). Large sample standard errors of kappa and weighted kappa. *Psychological Bulletin, 72,* 323–327.

Gardner, W. (1995). On the reliability of sequential data: Measurement, meaning, and correction. In J. M. Gottman (Ed.), *The analysis of change* (pp. 339–359). Hillsdale, NJ: Erlbaum.

Gart, J. J., & Zweifel, J. R. (1967). On the bias of various estimators of the logit and its variance with application to quantile bioassay. *Biometrika, 54,* 181–187.

Garvey, C. (1974). Some properties of social play. *Merrill-Palmer Quarterly, 20,* 163–180.

Garvey, C., & Berndt, R. (1977). The organization of pretend play. In *JSAS Catalog of Selected Documents in Psychology, 7,* 107. (Ms. No. 1589).

Good, P. (1994). *Permutation tests: A practical guide to resampling methods for testing hypotheses.* New York: Springer-Verlag.

Goodenough, F. L. (1928). Measuring behavior traits by means of repeated short samples. *Journal of Juvenile Research, 12,* 230–235.

Goodman, L. A. (1983). A note on a supposed criticism of an Anderson–Goodman test in Markov chain analysis. In S. Karlin, T. Amemiya, & L. A. Goodman (Eds.), *Studies in econometrics, time series, and multivariate statistics* (pp. 85–92). New York: Academic Press.

Gottman, J. M. (1979a). *Marital interaction: Experimental investigations:* New York: Academic Press.

Gottman, J. M. (1979b). Time-series analysis of continuous data in dyads. In M. E. Lamb, S. J. Sumoi, & G. R. Stephenson (Eds.), *Social interaction analysis: Methodological issues* (pp. 207–229). Madison: University of Wisconsin Press.

Gottman, J. M. (1980a). Analyzing for sequential connection and assessing interobserver reliability for the sequential analysis of observational data. *Behavioral Assessment, 2,* 361–368.

Gottman, J. M. (1980b). The consistency of nonverbal affect and affect reciprocity in marital interaction. *Journal of Consulting and Clinical Psychology, 48,* 711–717.

Gottman, J. M. (1981). *Time-series analysis: A comprehensive introduction for social scientists.* New York: Cambridge University Press.

Gottman, J. M. (1983). How children become friends. *Monographs of the Society for Research in Child Development, 48*(3, Serial No. 201).

Gottman, J. M. (1990). Chaos and regulated change in family interaction. In P. Cowan and E. M. Hetherington (Eds.), *New directions in family research: transition and change.* Hillsdale, NJ: Erlbaum.

Gottman, J. M., & Bakeman, R. (1979). The sequential analysis of observational data. In M. E. Lamb, S. J. Suomi, & G. R. Stephenson (Eds.), *Social interaction analysis: Methodological issues* (pp. 185–206). Madison: University of Wisconsin Press.

Gottman, J. M., & Levenson, R. W. (1992). Marital processes predictive of later dissolution: Behavior, physiology, and health. *Journal of Personality and Social Psychology, 63,* 221–233.

Gottman, J. M., Markman, H., & Notarius, C. (1977). The topography of marital conflict: A sequential analysis of verbal and nonverbal behavior. *Journal of Marriage and the Family, 39,* 461–477.

Gottman, J. M., & Parker, J. (Eds.) (1985). *Conversations of friends: Speculations on affective development.* New York: Cambridge University Press.

Gottman, J. M., & Ringland, J. T. (1981). The analysis of dominance and bidirectionality in social development. *Child Development, 52,* 393–412.

Gottman, J., Rose, F., & Mettetal, G. (1982). Time-series analysis of social interaction data. In T. Field & A. Fogel (Eds.), *Emotion and interactions* (pp. 261–289). Hillsdale, NJ: Lawrence Erlbaum.

Gottman, J. M., & Roy, A. K. (1990). *Sequential analysis: A guide for behavioral research.* New York: Cambridge University Press.

Haberman, S. J. (1977). Log-linear models and frequency tables with small expected cell counts. *Annals of Statistics, 5,* 1148–1169.

Haberman, S. J. (1978). *Analysis of qualitative data* (Vol. 1). New York: Academic Press.

Haberman, S. J. (1979). *Analysis of qualitative data* (Vol. 2). New York: Academic Press.

Hamilton, G. V. (1916). A study of perserverence reactions in primates and rodents. *Behavior Monographs, 3* (No. 2).

Hartmann, D. P. (1977). Considerations in the choice of interobserver reliability estimates. *Journal of Applied Behavior Analysis, 10,* 103–116.

Hartmann, D. P. (1983). Assessing the dependability of observational data. In D. P. Hartmann (Ed.)., *Using observers to study behavior: New directions for methodology of social and behavioral science* (No. 14, pp. 51–65). San Francisco: Jossey-Bass.

Hartup, W. W. (1979). Levels of analysis in the study of social interaction: An historical perspective. In M. E. Lamb, S. J. Suomi, & G. R. Stephenson (Eds.), *Social interaction analysis: Methodological issues* (pp. 11–32). Madison: University of Wisconsin Press.

Hays, W. L. (1963). *Statistics* (1st ed.). New York: Holt, Rinehart, & Winston.

Hollenbeck, A. R. (1978). Problems of reliability in observational research. In G. P. Sackett (Ed.), *Observing behavior* (Vol. 2): *Data collection and analysis methods* (pp. 79–98). Baltimore: University Park Press.

Hubert, L. (1977). Kappa revisited. *Psychological Bulletin, 84,* 289–297.

Jaffe, J., & Feldstein, S. (1970). *Rhythms of dialogue.* New York: Academic Press.

Johnson, S. M., & Bolstad, O. D. (1973). Methodological issues in naturalistic observation: Some problems and solutions for field research. In L. A. Hamerlynch, L. C. Handy, & E. J. Mash (Eds.), *Behavior change: Methodology, concepts, and practice* (pp. 7–67). Champaign, IL: Research Press.

Jones, R. R., Reid, J. B., & Patterson, G. R. (1975). Naturalistic observations in clinical assessment. In P. McReynolds (Ed.), *Advances in psychological assessment* (Vol. 3, pp. 42–95). San Francisco: Jossey-Bass.

Kemeny, J. G., Snell, J. L., & Thompson, G. L. (1974). *Introduction to finite mathematics.* Englewood Cliffs, NJ: Prentice-Hall.

Kennedy, J. J. (1983). *Analyzing qualitative data: Introductory log-linear analysis for behavioral research.* New York: Praeger.

Kennedy, J. J. (1992). *Analyzing qualitative data: Log-linear analysis for behavioral research* (2nd ed.). New York: Praeger.

Knoke, D., & Burke, P. J. (1980). *Log-linear models.* Newbury Park, CA: Sage.

Krokoff, L. (1983). The anatomy of negative affect in blue collar marriages. Unpublished doctoral dissertation, University of Illinois at Urbana-Champaign.

Landesman-Dwyer, S. (1975). *The baby behavior code (BBC): Scoring procedures and definitions.* Unpublished manuscript.

Losada, M., Sanchez, P., & Noble, E. E. (1990). Collaborative technology and group process feedback: Their impact on interactive sequences in meetings. *CSCW Proceedings,* 53–64.

Miller, G. A., & Frick, F. C. (1949). Statistical behavioristics and sequences of responses. *Psychological Review, 56,* 311–324.

Miller, R. G., Jr. (1966). *Simultaneous statistical inference.* New York: McGraw-Hill.

Morley, D. D. (1987). Revised lag sequential analysis. In M. L. McLaughlin (Ed.), *Communication year book* (Vol. 10, pp. 172–182). Beverly Hills, CA: Sage.

Morrison, P., & Morrison, E. (1961). *Charles Babbage and his calculating engines.* New York: Dover.

Noller, P. (1984). *Nonverbal communication and marital interaction.* Oxford: Pergamon Press.

Oud, J. H., & Sattler, J. M. (1984). Generalized kappa coefficient: A Microsoft BASIC program. *Behavior Research Methods, Instruments, and Computers, 16*, 481.

Overall, J. E. (1980). Continuity correction for Fisher's exact probability test. *Journal of Educational Statistics, 5*, 177– 190.

Parten, M. B. (1932). Social participation among preschool children. *Journal of Abnormal and Social Psychology, 27*, 243–269.

Patterson, G. R. (1982). *Coersive family process.* Eugene, OR: Castalia Press.

Patterson, G. R., & Moore, D. (1979). Interactive patterns as units of behavior. In M. E. Lamb, S. J. Sumoi, & G. R. Stephenson (Eds.), *Social interaction analysis: Methodological issues* (pp. 77–96). Madison: University of Wisconsin Press.

Pedhazur, E. J., & Schmelkin, L. P. (1991). *Measurement, design, and analysis: An integrated approach.* Hillsdale, NJ: Erlbaum.

Rechten, C., & Fernald, R. D. (1978). A sampled randomization test for examining single cells of behavioural transition matrices. *Behaviour, 69*, 217–227.

Reynolds, H. T. (1984). *Analysis of nominal data.* Beverly Hills, CA: Sage.

Rosenblum, L. (1978). The creation of a behavioral taxonomy. In G. P. Sackett (Ed.), *Observing behavior* (Vol. 2): *Data collection and analysis methods* (pp. 15–24). Baltimore: University Park Press.

Raush, H. L., Barry, W. A., Hertel, R. K., & Swain, M. A. (1974). *Communication, conflict, and marriage.* San Francisco: Jossey-Bass.

Sackett, G. P. (1974). *A nonparametric lag sequential analysis for studying dependency among responses in observational scoring systems.* Unpublished manuscript.

Sackett, G. P. (1978). Measurement in observational research. In G. P. Sackett (Ed.), *Observing behavior* (Vol. 2): *Data collection and analysis methods* (pp. 25–43). Baltimore: University Park Press.

Sackett, G. P. (1979). The lag sequential analysis of contingency and cyclicity in behavioral interaction research. In J. D. Osofsky (Ed.), *Handbook of infant development* (pp. 623– 649). New York: Wiley.

Sackett, G. P. (1980). Lag sequential analysis as a data reduction technique in social interaction research. In D. B. Sawin, R. C. Hawkins, L. O. Walker, & J. H. Penticuff (Eds.), *Exceptional infant* (Vol. 4): *Psychosocial risks in infant–environment transactions.* New York: Brunner/Mazel.

Shannon, C. E., & Weaver, W. (1949). *The mathematical theory of communication.* Urbana: University of Illinois Press.

Shotter, J. (1978). The cultural context of communication studies: Theoretical and methodological issues. In A. Lock (Ed.), *Action, gesture, and symbol: The emergence of language* (pp. 43–78). London: Academic Press.

Siegel, S. (1956). *Nonparametric statistics for the behavioral sciences.* New York: McGraw-Hill.

Smith, P. K. (1978). A longitudinal study of social participation in preschool children: Solitary and parallel play reexamined. *Developmental Psychology, 14*, 517–523.

Smith, P. K., & Connolly, K. J. (1972). Patterns of play and social interaction in preschool children. In N. Blurton Jones (Ed.), *Ethological studies of child behavior* (pp. 65–95). Cambridge: Cambridge University Press.

Soskin, W. F., & John, V. P. (1963). The study of spontaneous talk. In R. G. Barker (Ed.), *The stream of behavior: Explorations of its structure and content* (pp. 228–287). New York: Appleton-Century-Crofts.

Sroufe, L. A., & Waters, E. (1977). Attachment as an organizational construct. *Child Development, 48*, 1184–1199.

Stern, D.N. (1974). Mother and infant at play: The dyadic interaction involving facial, vocal, and gaze behaviors. In M. Lewis & L. A. Rosenblum (Eds.), *The effect of the infant on its caregiver* (pp. 187–213). New York: Wiley.

Suen, H. K. (1988). Agreement, reliability, accuracy, and validity: Toward a clarification. *Behavioral Assessment, 10*, 343–366.

Suomi, S. J. (1979). Levels of analysis for interactive data collected on monkeys living in complex social groups. In M. E. Lamb, S. J. Suomi, G. R. Stephenson (Eds.), *Social interaction analysis: Methodological issues* (pp. 119–135). Madison: University of Wisconsin Press.

Suomi, S. J., Mineka, S., & DeLizio, R. D. (1983). Short- and long-term effects of repetitive mother–infant separations on social development in rhesus monkeys. *Developmental Psychology, 19*, 770–786.

Taplin, P. S., & Reid, J. B. (1973). Effects of instructional set and experimenter influence on observer reliability. *Child Development, 44*, 547–554.

Tapp, J., & Walden, T. (1993). PROCODER: A professional tape control coding and analysis system for behavioral research using videotape. *Behavior Research Methods, Instruments, & Computers, 25*, 53–56.

Tronick, E. D., Als, H., & Brazelton, T. B. (1977). Mutuality in mother–infant interaction. *Journal of Communication, 27*, 74–79.

Tronick, E., Als, H., & Brazelton, T. B. (1980). Monadic phases: A structural descriptive analysis of infant–mother face to face interaction. *Merrill-Palmer Quarterly, 26*, 3–24.

Tuculescu, R. A., & Griswold, J. G. (1983). Prehatching interactions in domestic chickens. *Animal Behavior, 31*, 1– 10.

Uebersax, J. S. (1982). A generalized kappa coefficient. *Educational and Psychological Measurement, 42*, 181–183.

Upton, G. J. G. (1978). *The analysis of cross-tabulated data.* New York: Wiley.

Wallen, D., & Sykes, R. E. (1974). *Police IV: A code for the study of police–civilian interaction.* (Available from Minnesota Systems Research, 2412 University Ave., Minneapolis, MN 55414.)

Wampold, B. E. (1989). Kappa as a measure of pattern in sequential data. *Quality and Quantity, 23*, 171–187.

Wampold, B. E. (1992). The intensive examination of social interaction. In T. R. Kratochwill & J. R. Levin (Eds.), *Single-case research design and analysis: New directions for psychology and education* (pp. 93–131). Hillsdale, NJ: Erlbaum.

Wickens, T. D. (1989). *Multiway contingency tables analysis for the social sciences.* Hillsdale, NJ: Erlbaum.

Wickens, T. D. (1993). Analysis of contingency tables with between-subjects variability. *Psychological Bulletin, 113*, 191–204.

Wiggins, J. S. (1973). *Personality and prediction.* Reading, MA: Addison-Wesley.

Williams, E., & Gottman, J. (1981). *A user's guide to the Gottman–Williams time-series programs.* New York: Cambridge University Press.

Wolff, P. (1966). The causes, controls, and organization of the neonate. *Psychological Issues, 5* (whole No. 17).

Index

Printed in the United Kingdom
by Lightning Source UK Ltd.
118294UK00001BA/4